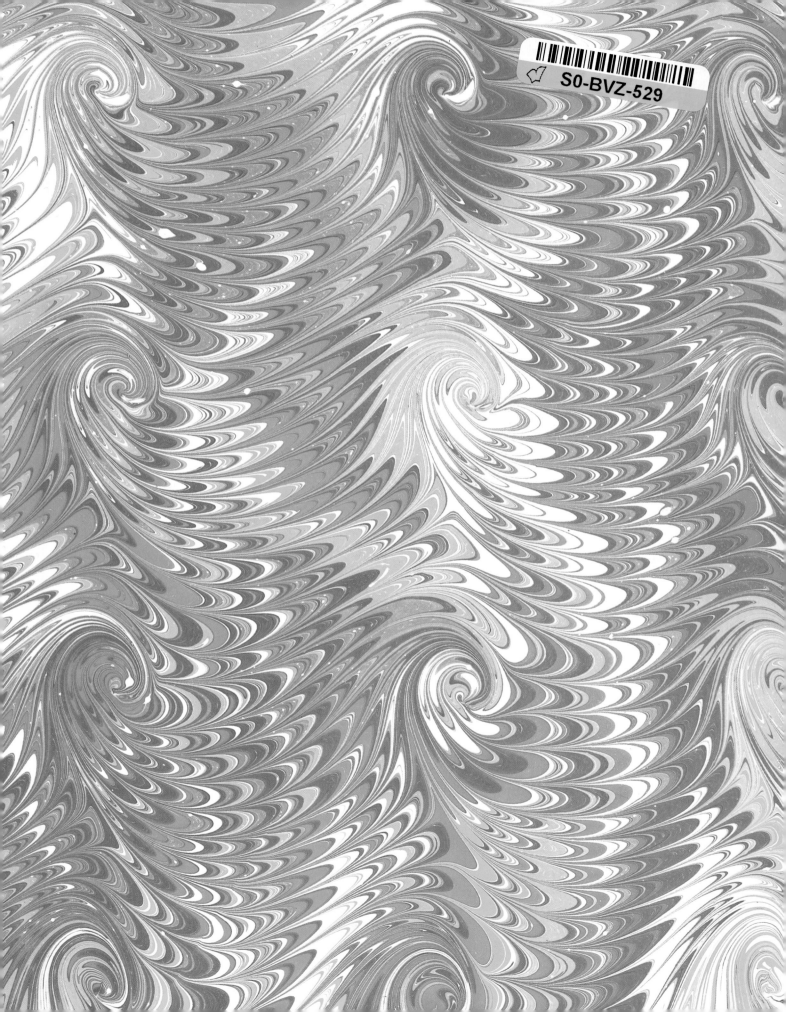

# Cigarette Lighters

Stuart Schneider & George Fischler

77 Lower Valley Road, Atglen, PA 19310

Library of Congress Cataloging-in-Publication Data

Schneider, Stuart L.
    Cigarette lighters / Stuart Schneider & George Fischler.
        p. cm.
    Includes bibliographical references and index.
    ISBN 0-88740-952-0
    1. Cigar lighters--Collectors and collecting--United States.
    2. Cigar lighters--United States--Catalogs.
    I. Fischler, George. II. Title.
TS2280.S36 1996
688'.4--dc20        95-26062
                CIP

Copyright© 1996 by Stuart Schneider & George Fischler

All rights reserved. No part of this work may be reproduced
or used in any form or by any means - graphic, electronic or
mechanical, including photocopying or information storage
and retrieval systems - without written permission
from the copyright holder.

This book is meant only for personal home use and
recreation. It is not intended for commercial
applications or manufacturing purposes.

Printed in China

ISBN: 0-88740-952-0

---

Published by Schiffer Publishing Ltd.
77 Lower Valley Road
Atglen, PA  19310
Please write for a free catalog.
This book may be purchased from the publisher.
Please include $2.95 postage.
Try your bookstore first.

# Contents

Preface ............................................................. 5
History .............................................................. 6
Collecting Lighters ......................................... 8
Valuing Lighters ............................................ 10
Lighter Companies: History & Comments ........... 11
Bibliography ................................................... 174
Resources ....................................................... 174
Index .............................................................. 175

# Acknowledgements

No man is an island, said John Donne. This is never truer than when you are trying to put together a book. One person's knowledge is helpful, two people's knowledge is wonderful, but three or more are needed to be interesting and authoritative. The authors wish to thank the following people who graciously agreed to let them photograph parts of their collections or contributed information and without whose help this book would not have been possible. Authorized Repair Service, Charles Cohn, Andrew Fingerman, Jeremy Fingerman, Eric Fingerman, International Lighter Collectors Club, Judd Perlson, Ira and Julie Pilosoff, Judith Sanders, Larry Tolkin, Stuart Unger, Richard Weinstein, Vintage Lighters, Inc. and Zippo Manufacturing Company. We really appreciate the extra help given by the Fingermans, the Pilosoffs, Richard Weinstein and Larry Tolkin.

These people's hospitality, desire to make this a great book, and depths of collection added significantly to the quality of this book.

# LIGHTER COMPANIES: HISTORY & COMMENTS

**ABDULLA:** Abdulla-made pipes, lighters and cigarettes during the 1920s to 1940s in France. They are extremely high quality lighters that used a unique mechanism. The company was purchased by the Querica family in the 1920s. Currently, they sell for $150 and up.

**ABDULLA,** ca.1934. A hard to find lighter with an Art Deco pattern. 2.25 inches tall. $125-$175

**ACS**, ca.1970. The American Cancer Society put out this "I quit smoking" "lighter". When your pushed down on the plunger, the little sign popped up. 1.75 inches tall. $5-$12

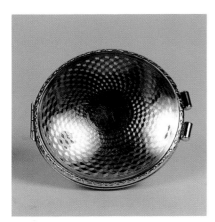

**ALLBRIGHT**, ca.1935. The Allbright "compact" lighter looked like a compact when closed. Made in USA. $100-$145

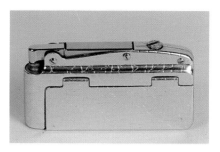

**ALMO,** ca.1956. The Swiss army knife of lighters. It came with a knife, file, screwdriver/bottle opener and a ballpoint pen. Made in the U.S. and 2.25 inches tall. $125-$145

**AMBASSADOR,** ca.1928. A high quality lift arm lighter. Made in the U.S. and 2 inches tall. $45-$75

**AMSCION,** ca.1920. Figural striker table lighter with a clock. 7 inches tall. $175-$225

**ARER,** ca.1929. A beautiful lift arm mother-of-pearl covered lighter/cigarette holder set in its original box. Made in Germany. $150-$200

**ASPREY,** ca.1939. Concealed mechanism, sterling silver lighter. Made in England and 2 inches tall. $750-$800

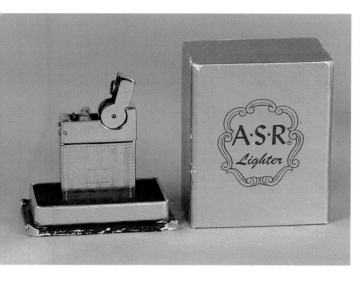

**ASR,** ca.1949. The ASR was made by the American Safety Razor Company. It was a good quality automatic pocket lighter. The company was based in Brooklyn, New York. $33-$45

**ASR,** ca.1950. A great boxed lighter with two additional leather covered sleeves to change the appearance of the lighter to match the user's attire. $70-$85

**ASR,** ca.1952. An unusual combination lighter and heart locket. Made in USA. $50-$75

**ASR.,** ca.1950. American Safety Razor pocket lighter with a plastic mother-of-pearl side panel. $30-$40

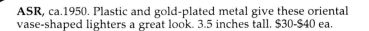

**ASR,** ca.1950. Plastic and gold-plated metal give these oriental vase-shaped lighters a great look. 3.5 inches tall. $30-$40 ea.

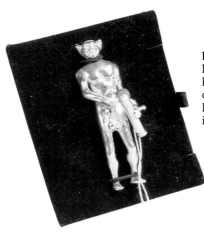

**BALL**, ca.1994. Richard Ball reproduced this early "Imp" lighter in sterling silver. The Imp's hand was moved to expose his penis tipped with a wick. The striker wheel, held in the other hand, showered him in sparks. This was a limited edition lighter, limited to 200 numbered pieces made in England. 2 inches tall.

**BAIER**, ca.1940. A very high quality, small rolling table lighter, ashtray and cigarette case. All metal (aluminum and brass) except for the tires. Made in Germany. $175-$200

**BAIER:** 1930s. Made in Germany during the 1930s and 1940s. The post war pieces are often marked "U.S. Zone". A high quality lighter that had a Ronson type mechanism.

**BAIER,** ca.1947. A small aluminum lighter in the shape of a Jeep- type vehicle. Made in the Germany and 2 inches tall. $150-$175

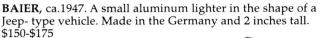

**BAIER,** ca.1947. This German lighter, cigarette box and ashtray was made of machined aluminum. Beautifully designed and functional, it was typical of high quality German craftsmanship. It is about 14 inches long. $175-$200

**BEATTIE JET:** 1945 to 1960s. An American-made high quality pipe lighter available with different coverings in leather, sterling silver, etc. Advertising of the period stated that not only could they light pipes, but they could do light soldering jobs and thaw frozen car locks. They are abundant in the United States and harder to find in Europe. Demand is higher in Europe.

**BEATTIE,** ca.1960. The Beattie Jet was an innovative lighter made for lighting pipes. When the standard wick would light and heat up the nozzle, a jet of air would blow the flame, like a small blow torch, into the bowl of the pipe. It used regular lighter fluid. Some pipe users claimed that it would scorch the pipe bowl. $25-$45

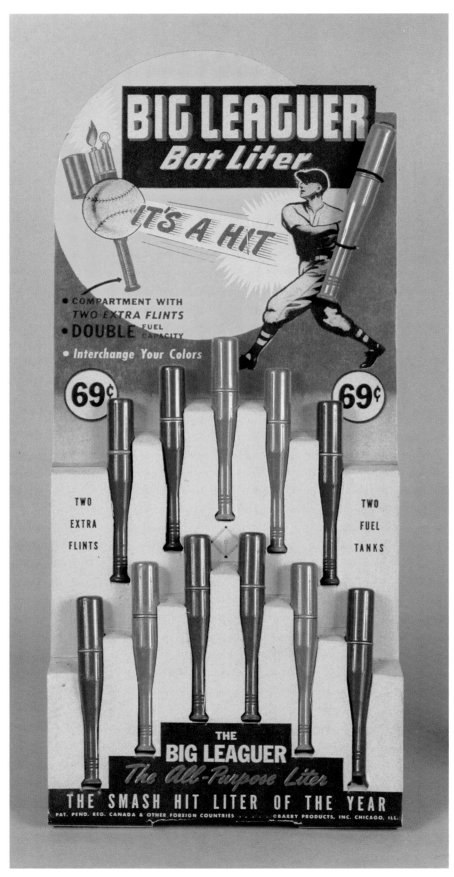

**BIG LEAGUER,** ca.1940s. A novelty lighter in the shape of a baseball bat, made in Chicago. $175-$200 for all

**R. BLACKINGTON CO.,** ca.1935. A high quality sterling silver pocket lighter. Made in the U.S. and 2.5 inches tall. $300-$325

**BRAUN,** ca.1970. The "toaster" butane table lighter with great engineering. It was one of the first. 3.5 inches tall. $40-$50

**K.W. BONET,** ca.1950. A large snakeskin covered table lighter with a semi-automatic action. $90-$105

**BOWERS,** ca.1938. The Bower's pipe lighter was ingeniously fitted with a striker mechanism built into the bowl end of the pipe. 7 inches long. Made in USA. $500-$600

**BRONICA,** ca.1969. A "World Clock" model butane table lighter inspired by the space program. When a lever is depressed, a battery powered coil heats up and lights the butane. 9 inches tall. $25-$40

**BRUMA**, ca.1925. Hand engraved sterling silver pocket lighter. Excellent detailing; notice the lizard on the windscreen. This very rare lighter has an unusual plunger mechanism.

**BULGARI**, ca.1990. Two extremely nice modern lighters, both in 18kt gold. The round lighter in the box is the "Firewheel" gas lighter, while the slightly leaning lighter is the "Turbogas" model than has a flexible spring-like body. 2.25 inches and 3.5 inches. $1000-$1600

**BUTABLOC**, ca.1950. French-made early gas pocket lighter, streamlined on the outside and complex on the inside. $30-$40

**CAPITAL**, ca.1912. A complex lever-actuated table lighter made in the U.S.A. of copper-plated brass. 5.25 inches tall. $150-$185

**CAPITAL**, ca.1912. A complex lever-actuated table lighter made in the U.S.A. of copper-plated brass. 5.25 inches tall. $175-$200

**CARLTON LIGHTERS:** 1918 to 1930s. The Carlton was a division of the Kum-a-part Company which made snap buttons for clothing. The mechanisms were high quality and the enamel overlay models are very desirable. All are very collectible. The company died with the Depression.

**CARLTON,** ca.1923. Simple but sturdy, this thin Carlton man's pocket lighter has a leather cover. $60-$75

**CARTIER,** ca.1930. A rather incredible cigarette dispenser/lighter made of sterling silver. The lever on the right is pushed down and automatically a cigarette falls down the ramp, the lighter lights and a bellows inhales air and the box dispenses a lit cigarette. 5.5 inches tall and 7 inches wide.

**CARTIER:** 1847 to Present. Cartier was founded in Paris, France by Louis Francois Cartier (1819-1904) who took over the shop of a jeweler named Picard. His work was of the highest caliber and he soon began to do work for King Louis Philippe and other royal and aristocrat families. His son, Alfred (1841-1925), joined him in 1872 and continued the tradition of high quality products. The firm moved to the Rue de la Paix in Paris in 1898. Alfred's son, Louis-Joseph took over control from his father around the turn of the century and began the detailed record keeping that is envied today. If you bought a Cartier product in 1900, Cartier has the records to show who made the product, when it was made, what it sold for and when and to whom it was sold. The firm was innovative. For example, the wristwatch was an invention of Louis-Joseph.

Lighter production was established after World War I. Lighters were produced before the war on an a per customer basis. Cartier produced wonderful lighters in precious metals with some containing precious gemstones. The demand increased for their lighters during the 1920s and, by the late 1920s, their products were popular with customers other than the wealthy. They used high quality production methods and their products are well-made and well-conceived. Cartier introduced a gas model in 1968 called the Ovale.

**CARTIER,** ca.1940. Cartier made very handsome lighters in gold and silver. This sterling silver model sports a hidden pocket hinge. The trim is 18kt gold. $400-$500

**CHASE,** ca.1937. A 3 inch tall ball lighter. Made in USA. $50-$70

**CLARK:** 1920s to 1940s. The company was located in Massachusetts and made a nice solid, high quality lighter available in many styles including gold, silver, with watches, etc. Currently there is a moderate demand for these lighters.

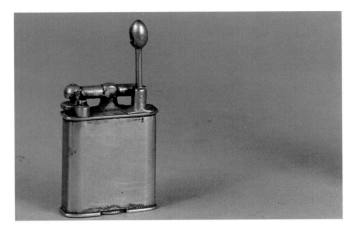

**CLARK,** ca.1924. A high quality man's or woman's pocket lighter. Made in USA. $35-$45

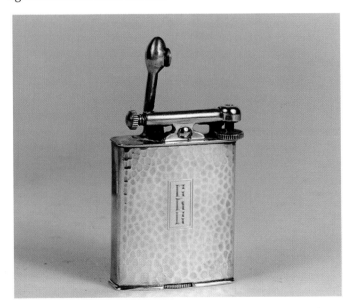

**CLARK,** ca.1928. The hammered sterling silver Clark "Firefly" lift arm lighter. Made in the U.S. and 2.25 inches tall. $250-$300

**CLARK,** ca.1940. A wood body pocket lighter. 2.25 inches tall. $80-$110

**CLARK,** ca.1940. A nice Clark pocket lighter with a leather cover and windscreen. 2.25 inches tall. $110-$130

**CLASSIC,** ca.1930. "Ball" model in a sterling silver. English-made. $600-$700

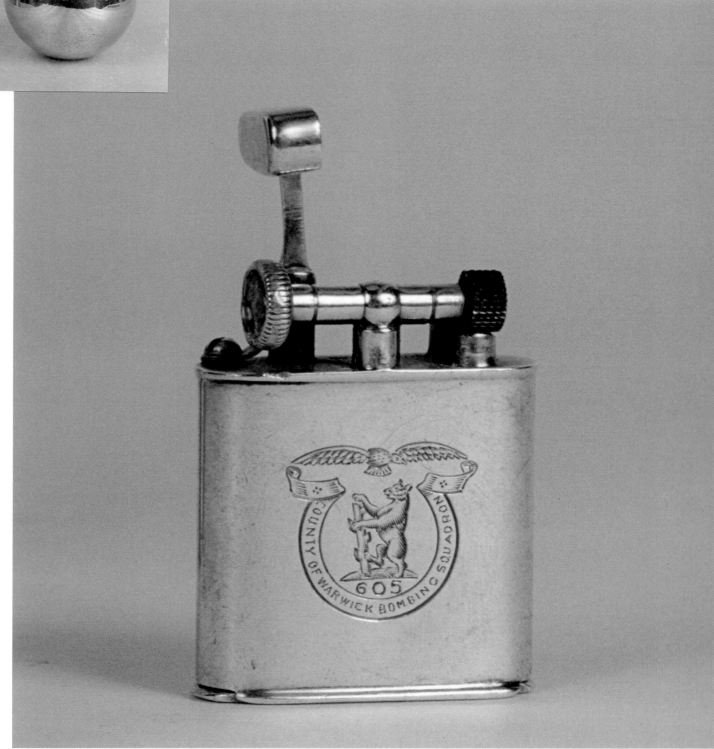

**CLASSIC,** ca.1939. A handsome man's pocket lighter made of sterling silver and marked with the "County of Warwick Bombing Squad". Made in England. $250-$275

**CLASSIC**, ca.1940. An English "Jumbo" model table lighter. This is unusual in that it operates by a wheel at the back of the lighter rather than at the sparking wheel area. 4 inches tall. $125-$150

**CLODION**, ca.1930. Several makers produced a lighter in the shape of a Jerry Can. This one was made in France. $130-$160

**COLBY**, ca.1939. Table lighter that works by pushing in one of the sides at the top. The central portion opens and lights. 3.75 inches tall. $145-$165

**COLIBRI**, ca.1929. A wonderful niello Art Deco lighter. When the side lever is pushed down, the top pops up. Releasing the lever creates the spark. This is Colibri's first lighter. $160-$180

**COLIBRI:** 1920s to present. Originally a German firm called JBELO, which stood for Julius & Ben Lowenthal, that made pipes and lighters. They were English made with current demand in the United States higher than England. In 1928, Julius developed the open spring mechanism on their "Original" model and Colibri introduced it. In 1933, the Colibri part moved to London while JBELO stayed in Germany. It is a high quality lighter. The company was dissolved in 1993.

**COLIBRI**, ca.1935. Colibri lighters are instantly recognizable by the outside lever. This model was called the "Original" and was made with a silver wrap over chrome-plated brass. It is 2 inches tall. $135-$165

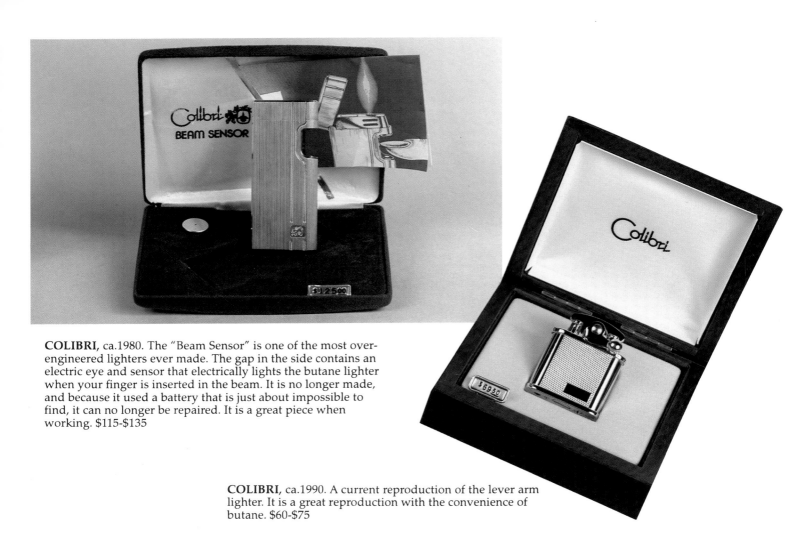

**COLIBRI,** ca.1980. The "Beam Sensor" is one of the most over-engineered lighters ever made. The gap in the side contains an electric eye and sensor that electrically lights the butane lighter when your finger is inserted in the beam. It is no longer made, and because it used a battery that is just about impossible to find, it can no longer be repaired. It is a great piece when working. $115-$135

**COLIBRI,** ca.1990. A current reproduction of the lever arm lighter. It is a great reproduction with the convenience of butane. $60-$75

**CHRISTIAN DIOR,** ca.1970. Cigarette case and table lighter in the shape of an artichoke. 5.25 inches tall. $125-$175

**DANDY**, ca.1920s. Austrian-made gun-shaped lighter. 2.75 inches long. $75-$100

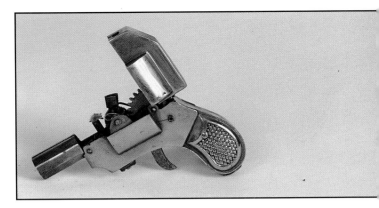

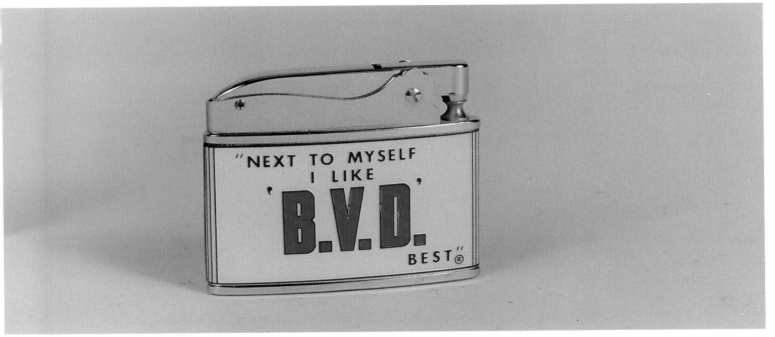

**DIRECT**, ca.1955. Two views of a painted chrome-plated lighter advertising underwear. Made in Japan and 2 inches tall. $50-$75

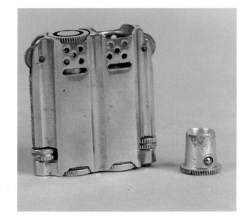

**DOUBLELITE**, ca.1945. An unusual pocket lighter, it is a Siamese twin. Two separate lighters joined in the middle. Made from machined aluminum and 2.25 inches tall. $175-$225

**DOUGLASS LIGHTERS:** 1920s. Douglass sold a high quality lighter. They were made in Menlo Park and sometimes marked "Wrigley Bldg, Chicago". They made lighters for Van Clef & Arpels and covered the market with everything from inexpensive to elegant, expensive products. Their semi-automatic swing arm action is an eye catcher.

**DOUGLASS**, ca.1926. A unusual California-made (Menlo Park) lighter with a semi-automatic action. The flint wheel is incorporated into the snuffer arm (missing snuffer) and a heavy spring action flips the snuffer arm back when a button is pressed, creating the spark. 2.25 inches tall. $75-$95

**DUNHILL:** 1907 to present. Dunhill quality is world famous and, as such, their lighters are actively sought throughout the world. This is reflected in the high prices paid for some of the rare models. Dunhill was founded by Alfred Dunhill in 1907 on Duke Street, London. Before 1907, Alfred continued a horse and carriage accessories shop founded by his father. With the advent of the automobile, the shop began to make quality items for the automobile driver. One successful item (1904) was a pipe with a built in windscreen. Alfred left the automobile accessories business and began a pipe, tobacco, and smoking accessories shop.

His first Alfred Dunhill pipe, made from the best briar, was introduced in 1910. The concept of the Dunhill shop was to offer the finest products. These cost more, but Dunhill wanted to compete with other tobacco shops not on price, but on quality.

Dunhill's first lighter was the "Ednite" introduced in 1914. It was a pocket tinderwick lighter. The "Tinderwick", pocket match style lighter, followed in 1916. With the advent of World War I, soldiers became the largest business segment of sales of pipes and tobacco. Dissatisfied with the lighters on the market, Alfred wanted to sell a lighter that was easy to use and could be lit with one hand.

The "Every Time" lighter was introduced in the Christmas 1923-1924 catalog. It was made for Dunhill by the English company of Wise and Greenwood. Within a year, the Every Time became the "Unique". Demand was so high that the Geneva firm of La Nationale was also engaged to make the lighters. One of the most sought after styles, a lighter with a watch, was not a Dunhill idea but a customer's idea. Santiago Soulas, a wealthy South American customer wanted a lighter with a watch in the side. Dunhill liked the idea and began having watch lighters made in 1926.

In the 1922/1923 season, Dunhill created the Parker Pipe Company to sell pipes and smoking accessories that were not perfect enough to be called Dunhill. (Think of this as the Bentley as not quite as good as a Rolls Royce). Parker Pipe was successful and it introduced a series of lighters sold under the name "Beacon".

Alfred Dunhill retired in 1928 and his Brother, Herbert, took over control of the company aided by Alfred's daughter, Mary.

The Namiki Manufacturing Company of Tokyo made a line of high quality products with hand painted finishes. They were looking for new markets and contacted Dunhill. Dunhill signed an exclusive agreement to sell the Dunhill-Namiki line of smoking accessories and writing instruments on July 8, 1930. On August 29, 1930, Dunhill signed an agreement with the American Safety Razor Company to market Dunhill lighters in America. Always looking to make his products better, an improved Unique model with a double wheel striker was introduced in 1931 along with a self-winding watch lighter. A minor problem for collectors is that many customers returned their single wheel lighters to Dunhill for updating to double wheel lighters. This makes the single wheel lighters rarer and could be confusing when dating an early double wheel lighter that has been modified by Dunhill.

The first (1933) Dunhill "Tallboy" lighter was unusual as it has both the Dunhill and Cartier names on it. Cartier held the patent and Dunhill held the license to produce and sell it. Dunhill's Hunting Horn lighter was introduced in 1934. Its "Broadboy" line first appeared in 1936 along with a giant Unique lighter. 1938 saw the first Tinder Pistol Lighter and the Broadboy watch lighter come out in 1939.

Lighters were selling briskly and when the first gas lighter, called the Gentry, was introduced in Paris in 1947, Dunhill began working on its own gas lighter. Their first model, the "Rollagas" came out in 1956.

The Dunhill "Rollalite," an product exported only to Europe, came out in 1948. The "Sylph" first appeared in 1953, the same year as the "Aquarium" (Dunhill's records may not be accurate, as the Aquarium is rumored to appear in a Christmas, 1951 catalog). The "Sylphide" came out in 1958 and the gas Sylphide model followed in 1965.

**DUNHILL**, ca.1926. One of the most elegant lighters made. It is 14kt gold with a watch set in the body. The front opens from the top outward for access to the watch.

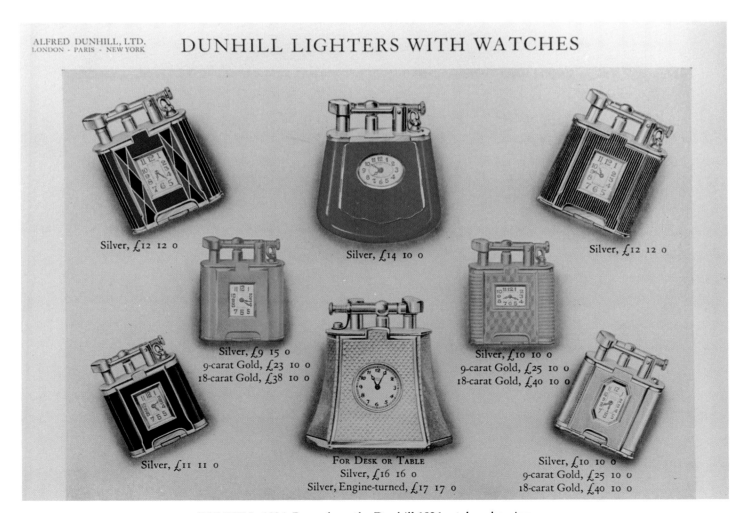

**DUNHILL**, 1926. Pages from the Dunhill 1926 catalog showing enamel lighters.

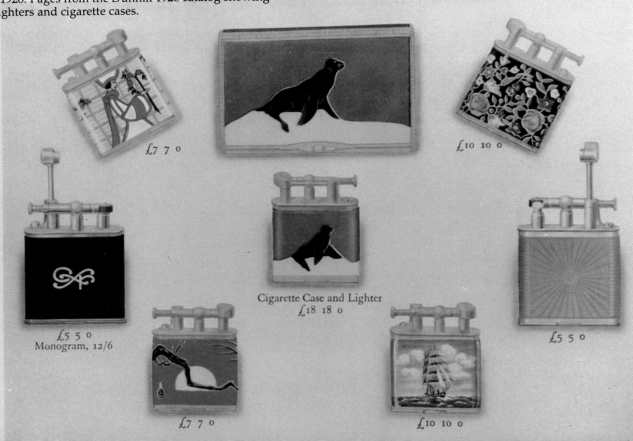

**DUNHILL**, 1926. Pages from the Dunhill 1926 catalog showing decorative lighters and cigarette cases.

**DUNHILL**, ca.1928. The Dunhill "Club" is an elegant sterling silver table lighter with classic, engine turned styling. 6 inches tall. $750-$800

**DUNHILL**, ca.1928. Dunhill "Unique" lighters in enameled sterling silver with Art Deco engine turned designs. The narrow-band model was made in Switzerland and the wide band model was made in Germany. 2.25 inches tall. $1000-$1200 each.

**DUNHILL**, ca.1928. Dunhill "Unique" lighter in enameled sterling silver. Made in England and 2.25 inches tall. $1000-$1200

**DUNHILL**, ca.1928. Dunhill "Unique" lighter in sterling silver with inset watch. Made in Switzerland and 2.25 inches tall. $1500-$1700

**DUNHILL**, ca.1928. Dunhill "Unique" lighter in two tone woven 14kt gold metal. Made in the U.S. and 2.25 inches tall. $1500-$1700

**DUNHILL**, ca.1929. Matching dueling pistols with fitted box. Each gun is about 3.5 inches tall. When the trigger is pulled, the top opens and the striker mechanism lights the wick. The hammer has been wired back to allow the mechanism to show. $225-$275 each gun alone.

**DUNHILL**, ca.1930. "Ball" model in a gold-plated metal. It was advertised that it would always stand upright however placed down.

**DUNHILL**, ca.1933. A classic Dunhill lighter in two tone 14kt gold with Art Deco engine-turned design. This was retailed by Cartier and is so marked. Made in the U.S. and 2.25 inches tall. $1500-$1700

**DUNHILL**, ca.1932. Silver-plated "Unique" table lighter. 5 inches tall. $300-$350

**DUNHILL**, ca.1932. One of the classics that might be found in an advanced collection, is the Dunhill sterling silver, large sized "Sport" lighter/watch combination in its original box. French-made and about 2.5 inches high. $2000-$2400

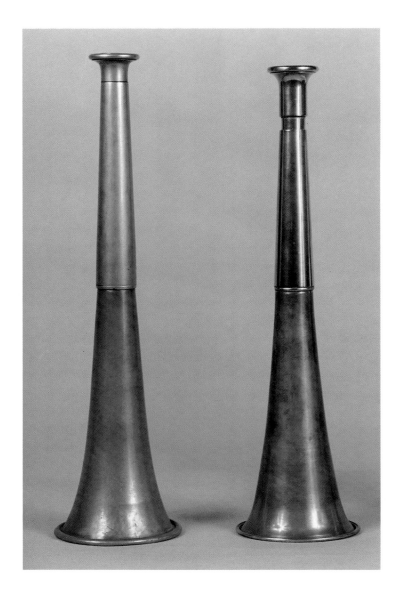

**DUNHILL**, ca.1936. "Hunting Horn" models of brass. The horn on the left is activated by pressing down on the "mouthpiece" end. It is extinguished by pushing the button on the small end. The horn on the right is activated by lifting and turning it lighter side up. An interior weight drops down and lights the lighter. It is extinguished manually by closing the snuffer.

**DUNHILL,** ca.1936. Dunhill "Unique" lighters in enameled sterling silver with Art Deco designs. Made in France and 2.25 inches tall. $1250-$1500 each.

**DUNHILL,** ca.1939. A "Unique" sports lighter in leather-covered silver plate with a wind screen. This is the Bijou or small lady's model. 2 inches tall. $200-$250

**DUNHILL,** ca.1943. A "Classic" ball lighter in silver plate. 2 inches tall. $500-$600

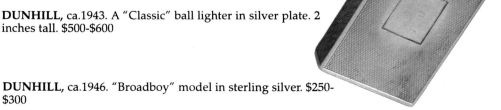

**DUNHILL,** ca.1946. "Broadboy" model in sterling silver. $250-$300

DUNHILL, ca.1946. "Salaam" model in black anodized aluminum. An unusual feature is that the thumbwheel has a spare striker wheel on the bottom when removed.

DUNHILL, ca.1949. "Silent Flame" model table and cigarette case model in Bakelite. This is a difficult model to find. $150-$200

DUNHILL, ca.1946. "Corona" model in a black enamel with gold filled trim. This lights automatically. Made in the U.S.

DUNHILL, ca.1949. "Silent Flame" model table model with ashtray. This is also a difficult model to find. $175-$225

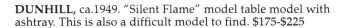

DUNHILL, ca.1950. A "Unique" lighter with a pipe lighting extension in silver plate. 2.5 inches tall. $225-$260

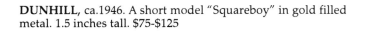

DUNHILL, ca.1946. A short model "Squareboy" in gold filled metal. 1.5 inches tall. $75-$125

**DUNHILL**, ca.1950. "Silent Flame" models used a battery to heat a coiled wire next to the wick when the metal wand was touched to the wire "fence" and the center ornament. 4 inches across. These are the "San Francisco Sun Tower," the "Ball Dancer," and the "Bulldog." $125-$225

**DUNHILL, ca.1950.** Carlton blue ceramic with spider web design about 3 inches long. An exceptionally nice table lighter in silver plate.

**DUNHILL, ca.1950.** "Silent Flame" models used a battery to create a spark when the metal wand was touched to the wire "fence" and the center ornament. 4 inches across. The right model is the "Sailboat" and the left is the "Sally Rand - Fan Dancer". L-R $65-$80, $120-$130

**DUNHILL, 1950.** Page from the 1950 Dunhill export catalog showing the Sports and Pipe lighters.

*Page 31*

# DUNHILL LIGHTERS
## MADE IN ENGLAND

ALFRED DUNHILL LTD
LONDON - PARIS - NEW YORK

### THE "SPORTS" LIGHTER
Regd. Design No. 737418.

The windscreen ensures a light which is easily maintained under all weather conditions. Silver Plated and Gold Plated, Plain and Engine Turned finish.

S'allume et résiste parfaitement au vent. Recommandé pour toutes les activités de la vie au grand air. Argenté ou doré, uni ou guilloché.

El abrigo contro el viento garantiza ignición infalible y una llama que se mantiene aún en las peores condiciones atmosféricas. Plateado y dorado,

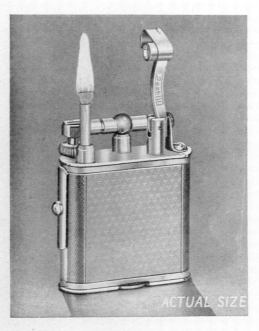

### THE "PIPE" LIGHTER
Regd. Design No. 737418.

Fitted with a wick which can be raised to afford an easy light for the pipe. Silver Plated and Gold Plated Plain and Engine Turned finish.

Muni d'un dispositif très simple permettant de relever le porte-mèche pour faciliter l'allumage de la pipe. Argenté ou doré, uni ou guilloché.

Posea un dispositivo que permite levantar la mecha, muy práctico para la

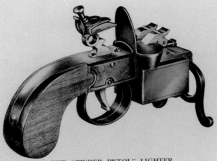

**DUNHILL, 1950.** Page from the 1950 Dunhill export catalog showing the Tinder Pistol lighter.

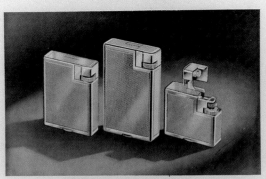

**DUNHILL, 1950.** Page from the 1950 Dunhill export catalog showing the Handy, Long Handy and Squareboy lighters.

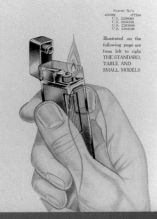

**DUNHILL, 1950.** Page from the 1950 Dunhill export catalog showing the Auto-Rollalite lighter.

**DUNHILL,** ca.1960. Tinderbox pistol. The gun is about 3.5 inches tall. When the trigger is pulled, the front, top opens and the striker mechanism lights the wick. The Dunhill gun lighters were great examples of novelty table lighters. Workmanship is top quality. Produced in the U.S. and England. There is a much rarer gas version produced in the mid-1960s. $60-$90

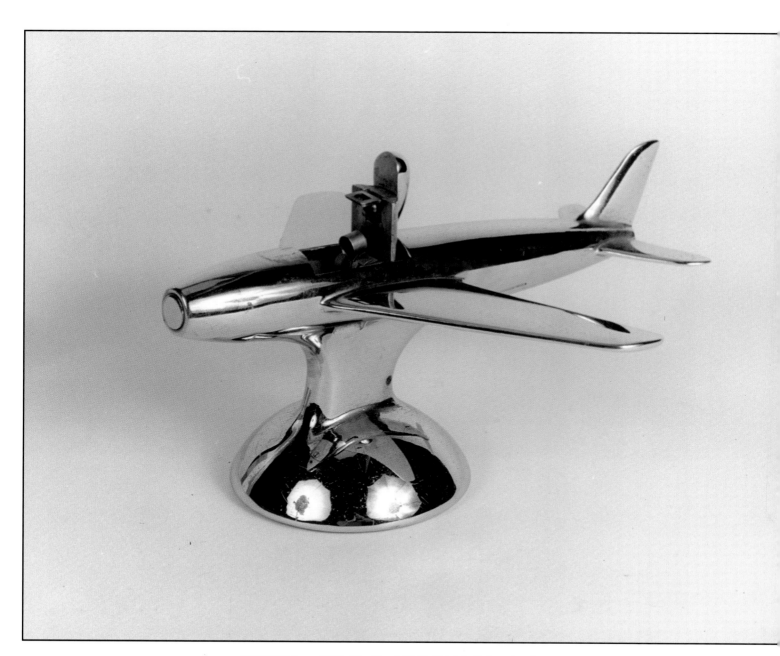

**DUNHILL,** ca.1952. The Dunhill "Jet" table lighter in chrome-plated metal. Push in the nose and the cockpit lights up. 4.5 inches tall. $650-$700

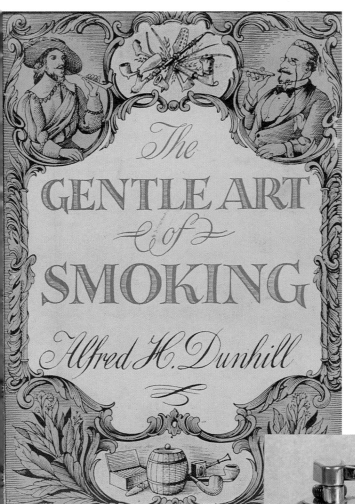

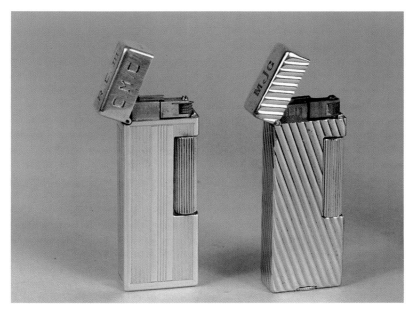

**DUNHILL**, ca.1955. Two "Rollalites" in sterling silver. Tall and thin, these were very popular lighters. The left one has an automatic action that pops the top when the wheel is turned, while the one on the right works by opening the lid and then turning the wheel. $125-$175

**DUNHILL**, 1954. The book *The Gentle Art of Smoking* by Alfred Dunhill.

**DUNHILL**, ca.1955. "Aquarium" model. 4 inches long. This style of lighter was also available in other motifs, such as "Aviary". $700-$900

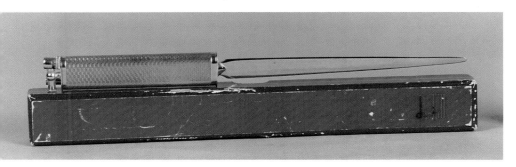

**DUNHILL**, ca.1956. "Sylph" model combination lighter/letter opener in gold filled metal. 9 inches long. $400-$500

**DUNHILL**, ca.1960. "Rollagas Rulerlite" model combination lighter/ruler in gold filled metal. 12 inches long. $350-$400

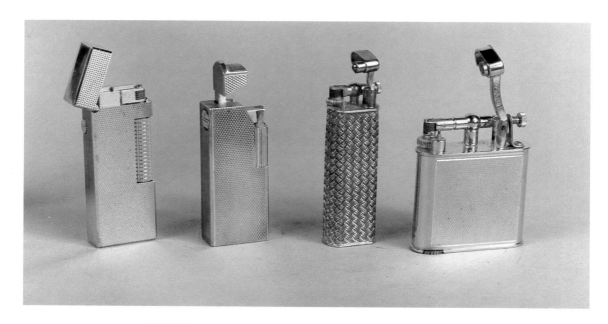

**DUNHILL**, Four Dunhill butane lighters in silver and gold filled metals. From left to right, the styles progress from the 1960s to 1970. A Swiss-made Rollagas model (still made today), a French "Aldunil", a French-made "Sylphide" and an English "Unique". $125-$350

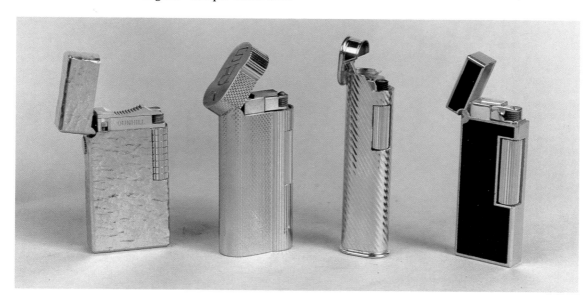

**DUNHILL**, Four Dunhill butane lighters in enamel and gold filled metal. From left to right, the styles progress from 1970 to 1990. An English-made "70", a Swiss-made "S", Swiss-made "Dress", and a Swiss-made "Gemline". $125-$175

**DUNHILL.** Two Dunhill butane lighters in gold filled metal from the 1990s. Both are English-made and are the "Unique" model and the "Sylphide" model. $125-$175

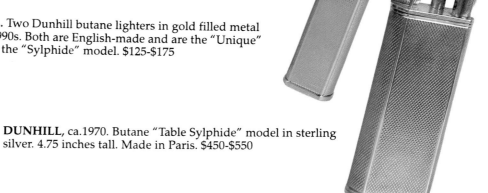

**DUNHILL**, ca.1970. Butane "Table Sylphide" model in sterling silver. 4.75 inches tall. Made in Paris. $450-$550

**S.T. DUPONT:** 1940. This French company was already well-established when they began to make and sell lighters in 1940. Their first lighters were brass models, but with the war shortages, other metals were used. Dupont introduced a gas lighter in 1953. The bottom was marked "HH" followed by 4 digits. In 1954, the bottom was marked "KK" followed by 4 digits. In 1955 and after, the code changed to "A" followed by 4 digits, then "B" followed by 4 digits, then "C" and the 4 digits, etc. They are known for their high quality gas models.

**S.T. DUPONT,** 1989. This butane lighter was made in honor of the French Revolution Bicentennial. $375-$450

**ELGIN-OTIS:** 1920s to 1940s. Elgin was the watch company and Otis was the lighter company. The earlier lighters marked "Elgin" are more desirable than those marked "Elgin-Otis". Nice quality.

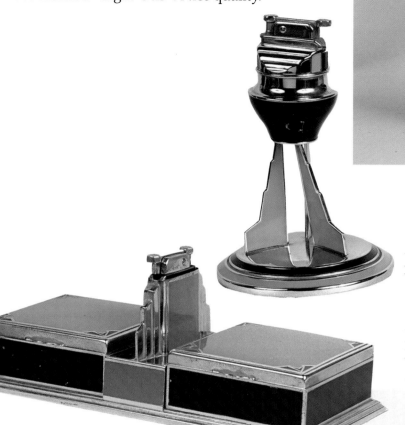

**ELGIN OTIS**, ca.1928. A sterling silver lighter with engine turning. 2 inches tall. $550-$700

**ELGIN-OTIS**, ca.1934. An elegant Art Deco-styled table top lighter. 7 inches tall. $500-$600

**ELGIN-OTIS**, ca.1935. An unusual Art Deco-styled table top lighter and double cigarette case. The unusual feature is that the lighter sits in a sleeve and is removable and usable as a pocket lighter. 12 inches wide. $250-$350

**ELGIN-OTIS**, ca.1935. A handsome Art Deco enameled lighter with matching cigarette case. Made in USA. $250-$325

**ELGIN-OTIS**, ca.1935. Deco-style table lighter 4.5 inches tall. Black enamel and chrome. $250-$350

**ETERNA,** ca.1936. Enameled sterling silver watch lighter. Made in Switzerland and 2.5 inches tall. $1500-$1700

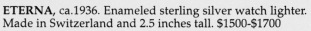

**EMSON,** ca.1936. A table top lighter with an unusual mechanism on a frosted glass base. 3.5 inches tall and made in Buffalo, NY. $75-$100

**EVANS:** 1920s -1960s. About 1917, Evans was manufacturing handbags in Massachusetts. They began to include handbag accessories and by the 1920s were manufacturing cigarette lighters. The lighters are often distinctive in shape and are of good quality, somewhat like a Ronson. There are some gas models made at the end of the line. The Art Deco models are especially nice and collectible and their 1950s models are always popular.

**EVANS,** ca.1926. Silver-plated metal lighter in the shape of a pitcher about 5.5 inches tall. The bottom opened and held cigarettes. This has the lift arm found on the first model Evans. Made in Massachusetts. $125-$150

**EVANS,** ca.1932. A beautiful mother-of-pearl covered lighter. Made in USA. $145-$185

**EVANS,** ca.1935. A very nice Evans pocket lighter. The body is sterling silver and the simulated snakeskin finish is hand-done enamel work. 2.25 inches tall. $250-$300

**EVANS,** ca.1932. The Evans "Triggerlite", a Deco sterling silver enameled lighter. Made in USA. $300-$350

**EVANS**, ca.1933. A handsome gold-plated Art Deco-styled combination cigarette case with lighter and watch. 7.5 inches tall. $125-$150

**EVANS**, ca.1935. Bakelite cased table lighter. 2.5 inches tall. Made in USA. $60-$85

**EVANS**, ca.1935. Chrome ball style table lighter. 3.5 inches tall. Made in USA. $75-$100

**EVANS**, ca.1935. A handsome Art Deco enameled lighter with matching cigarette case. Made in USA. $125-$170

**EVANS**, ca.1935. An Art Deco enameled combination lighter/ cigarette case. Made in USA. $110-$145

**EVANS**, ca.1935. This is the smallest case lighter made by Evans. Made in the U.S. and 2.75 inches tall. $125-$160

**EVANS**, ca.1935. A Bakelite Art Deco style lighter with a built-in clock. This is an exceptional piece in its masterful use of Bakelite, color, time, and fire. Made in the U.S. and 6 inches tall. $750-$800

**EVANS,** ca.1936. Elephant skin covered pocket lighter.

**EVANS**, ca.1938. Ball style table lighter. 4.25 inches tall. $75-$80

EVANS, ca.1940. A two-piece cigarette holder and lighter set. $150-$175

**EVANS**, ca.1940. An unusual wheel style lighter in an enameled case. 2.25 inches tall. $90-$110

**EVANS**, ca.1942. A lady's purse lighter with rhinestone decoration. 1.5 inches tall. $25-$35

EVANS, ca.1942. "Faberge" egg style table lighters. 4 inches tall. $70-$85 each.

**EVANS**, ca.1949. Mug lighter. It is a part of a series of mug lighters and is 5 inches tall. $25-$35

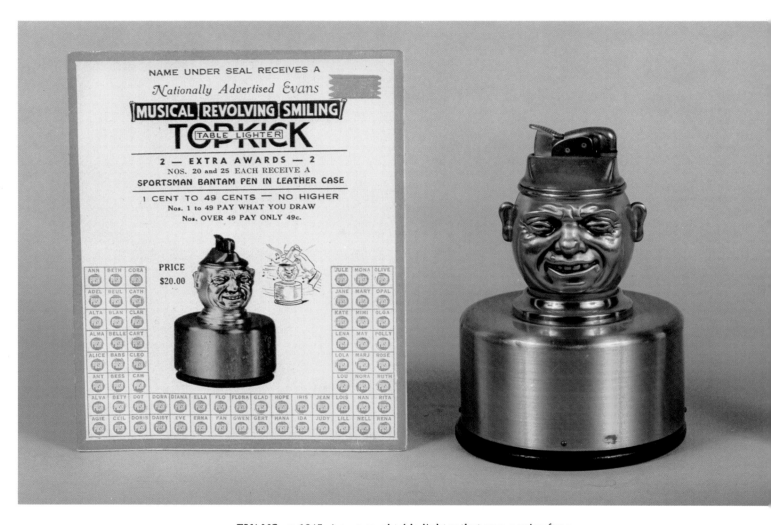

**EVANS,** ca.1945. An unusual table lighter that was a prize for a punch board game. When the lighter portion was removed, the face was a revolving music box. They call that a "smile." $75-$90 lighter alone.

**EVANS,** ca.1950. Two semi-automatic lighters in gold-plated metal. As the lighter is opened, the wheel is turned to produce the spark. The model on the left is the "Esquire" and it works by pressing down on the lid. The one on the right is the "Clipper" that lights by sliding a ridged piece (by the tiny screw) to the right. $50-$75

**EVANS,** ca.1950. Table lighter and cigarette case in a golden enamel finish. Lighter is about 5 inches tall. $60-$75

**EVANS,** ca.1952. "Spitfire" large pocket model lighters. The model above on the right is Bakelite and the one below on the right has a green crackle finish. Both have a moveable windguard. $35-$45

**EVANS,** ca.1953. A late Art Deco style lighter. Made in the U.S. and 2.25 inches tall. $25-$35

**EVANS**, ca.1953. This is a nice three-piece ashtray, cigarette holder and lighter set. $40-$60

**EVANS**, ca.1955. An unusually handsome pair of small pocket lighters to fit a lady's handbag. $25-$35

**EVERFLOW**, ca.1935. Bakelite table lighter. 4.5 inches tall. Made in USA. $60-$90

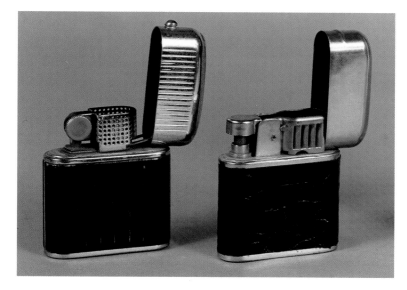

**EVANS**, ca.1955. Two substantial pocket lighters designed for men. The one on the right, works by depressing a plunger to rotate the spark wheel. The other works by turning a thumbwheel that turns the spark wheel. $50-$75

**EVER READY,** ca.1905. This Ever Ready "Vienna Plate" is an early colorful advertisement for the Ever Ready Cigar lighter.

**EVER READY,** ca.1905. Ever Ready is best known for its flashlights, but during its early years, it made and sold many products that used batteries. In 1906, it changed its name from American Electrical Novelty & Manufacturing Co. to American Ever Ready Co.

---

"EVER READY"
PORTABLE
ELECTRIC
CIGAR LIGHTERS

PATENTED
AND OTHER PATENTS PENDING

AMERICAN ELECTRICAL
NOV. & MFG. CO.
304-322 HUDSON ST., NEW YORK
CHICAGO, 184 LAKE ST.
BOSTON, 116 BEDFORD ST.
SAN FRANCISCO, 755 FOLSOM ST.
Address Nearest Office

---

"EVER READY"

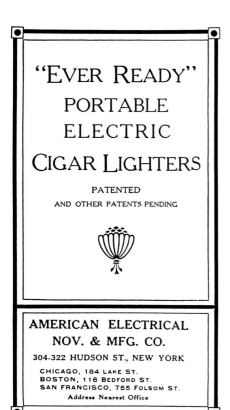

NO. 1837

¶ A very handsome floor stand twenty seven inches high. Especially designed for handy use in the club, card room and home. Equipped with an ash receiver with removable glass well and two cigar rests.

¶ Whether at the card table or reading in your easy chair there is always a light, a place for your ashes and a rest for your cigar or cigarette literally at your elbow. Heavily Nickel Plated.

Price, Complete, $10.00
Extra Battery, No. 326 . . .20

---

"EVER READY"

NO. 1840

¶ Neat, plain design of Brass, Nickel Plated and polished. Has removable Metal tray as illustrated.
Price, Complete, $3.50
Extra Battery No. 326, 20c.

NO. 1838 Same as above but without tray. Same battery. Price, Complete, $2.50

NO. 1841
¶ Beautifully Finished in Nickel Plate, heavily embossed with ornamental grape-vine pattern. Furnished with removable metal ash tray similar to No. 1840.
Price, Complete, $4.00
Extra Battery, No. 326 .20

NO. 1839
¶ Same of No. 1841 but without tray. Same battery.
Price, Complete, $3.00

---

PERFECTION

¶ From the Indian's two dry sticks rubbed together and Sir Walter Raleigh's flint and steel, down to the early friction sulphur matches and the modern parlor match there has been constant improvement in the manner of igniting the fragrant weed.

¶ When cigars first came into use an instant demand was created for a decorative, durable, inexpensive lighter to stand on library table, desk, cigar counter or bar and to be ever ready with a light. We have all seen the crude attempts to meet this demand with smoky, explosive kerosene torches, and the various lighters using gasolene all of which were unsatisfactory and objectionable for obvious reasons.

AT LAST PERFECTION IS REACHED
IN THE
"EVER READY"
Electric Cigar Lighter

¶ The only clean, smokeless, odorless and self-contained cigar lighter ever invented. The spark is supplied by a dry battery which will last for months with ordinary use and can easily be renewed. Smokeless, odorless and inexpensive wood alcohol is used in the reservoir which should be kept one quarter filled with fresh liquid.

¶ TO OBTAIN A LIGHT simply remove the torch from the reservoir and press the felt end down upon the perforated tip until the platinum coil glows. The alcohol will instantly ignite.

NO SMOKE      NO DIRT
NO ODOR       NO WIRES

---

"EVER READY"

NO. 1836

¶ This very attractive lighter is equipped, like No. 1837 on the previous page, with ash receiver and two cigar rests but instead of the floor stand has a strong clip for fastening on the edge of desk, card table or library table.

¶ This lighter should especially appeal to proprietors of cafes and restaurants as it can be securely clipped on the dinner table and effectually does away with dirt and danger from carelessly thrown matches, cigar and cigarette stubs. Cheaper to use than matches.

Price, Complete, $5.00
Extra Battery, No. 326, . .20

---

"EVER READY"

NO. 1848
¶ A Perfect Miniature Wooden Barrel about six inches high, furnished in Natural Maple or Mahogany Finish, highly polished. Trimmings Heavily Nickel Plated.
Price, Complete, $3.00
Extra Battery, No. 328 . . . .25

NO. 1847
¶ Same as above but barrel only four and one quarter inches high.
Price, Complete, $2.50
Extra Battery, No. 329, .20

**FIRECHIEF,** ca.1935. A Bakelite striker lighter. Made in the U.S. and 2.75 inches tall. $80-$100

**FLAMIDOR,** ca.1948. A "Favori" aluminum lipstick-shaped lighter that pulls apart in the middle to light. French-made and 3 inches tall. $95-$110

**FLAMINAIRE:** A French firm that made the first gas lighter. Manufacturing rights were owned by the Quercia family who owned the Flamidor company that had been making lighters in the several years just after the turn of the century. Marcel Quercia improved the model invented by Henri Pingeot in 1936. The first commercial gas lighter, the Gentry, a table model, was introduced in June, 1947. It was a very expensive lighter when first introduced. In 1948, the Crillon, a pocket model was introduced. The company did well, as the lighter was well-received and it was also the only maker of the gas refill canisters. The first automatic gas lighter was the Galet that came out in 1959

**FLAMINAIRE,** ca.1950. French-made butane table lighter. This is one of the first butane table lighters which used a disposable tank. The entire tank was replaced when empty. $50-$60

**Fluid Cans**, ca.1930s. As an adjunct to collecting lighters, many collectors are looking for the lighter fluid cans of the period of their lighters. $25-$50

**Fluid Cans**, ca.1940s. A nice selection of lighter fluid cans. $10-$15

**FOLMER & SCHWING**, ca.1892. A cased pocket watch lighter called a "Watch Pocket Lamp". It is early and very well-made. Folmer & Schwing made metal parts for several different industries. They are best known for their cameras and metal camera parts. American-made. Extra rare in the box.

**GARCIA GRANDE,** ca.1935. A store or bar cigar point of sale holder and lighter that was plugged into an electric outlet. Pushing the switch caused the heating element on the face plate to get hot. About 10 inches across. $200-$225

**GOLDEN WHEEL:** 1920s to about 1940. Golden Wheel made a very broad range of lighters, from the cheap trinket to high quality lighters. Most were medium quality. They are not too actively sought, but collector demand is growing.

**GOLDEN WHEEL**, ca.1934. Two handsome lighters from this high quality company, one in chrome and the other with a snakeskin cover. 2.5 inches tall. $75-$110

**GOLDEN WHEEL**, ca.1922. Man's pocket swing arm leather covered lighter. Made in the U.S. $35-$45

**GOLDEN WHEEL**, ca.1936. A geometric Art Deco design would have made this lighter a handsome showpiece when it was used to light a cigarette. 2.25 inches tall. $75-$125

**GOLDEN WHEEL**, ca.1934. A beautiful lighter with a built-in watch (high quality Cyma movement) in silver plate. 2.5 inches tall. $175-$225

**GOLDEN WHEEL**, ca.1924. Man's pocket lighter. Golden Wheel lighters are good quality and American-made. $35-$45

**GOLDEN WHEEL**, ca.1930. A public lighter fluid dispenser that would most likely have stood on a counter at a cigarette, cigar and pipe store.

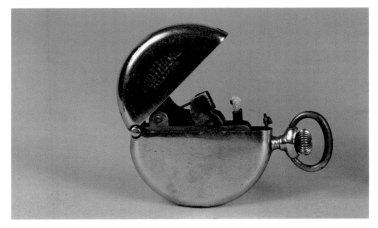

**GOTHAM**, ca.1916. A nice early automatic pocket lighter in the shape of a pocket watch. Pushing down on the stem releases a catch that lets the lid open and create the spark. $75-$100

**HARVEY AVEDON**, ca.1938. A high quality sterling silver pocket lighter. Made in the U.S. and 2.5 inches tall. $95-$140

**GUINN**, ca.1920. Bell-shaped countertop lighters. Heavy duty lighters for use in public places. 5 inches and 8 inches tall. L-R $65-$80, $150-$175

**H & H MFG.**, ca.1938. A convertible table lighter made of aluminum. It can be unscrewed from the base and a separate plug would screw in, converting it to a pocket lighter. 3.75 inches tall. $50-$75

**HARVEY AVEDON**, ca.1938. A handsome sterling silver table lighter. Made in the U.S. and 4.5 inches tall. $150-$185

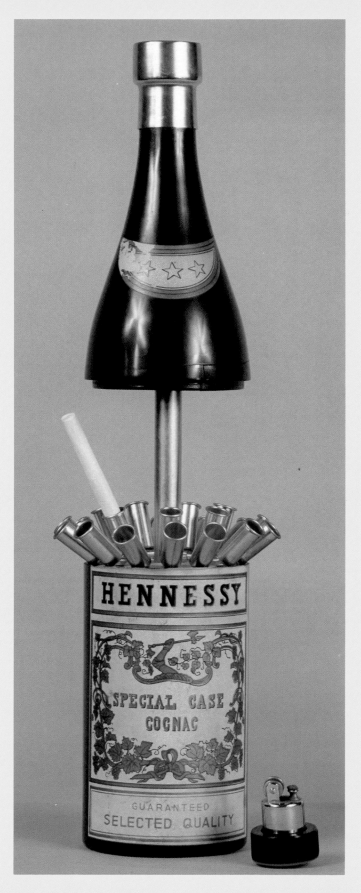

**HENNESSY,** ca.1955. A novelty cigarette holder and lighter. In the 1950s, it was popular to build a basement bar in your home. This was meant to sit on the bar and played music when opened. 15 inches tall. $75-$125

**MARATHON:** 1915 to 1940s. Marathon lighters are highly sought after, being of very high quality. The earliest ones (pre-1925) are highly prized as are their cigarette case/lighter combinations.

**MAGIC CASE,** ca.1933. Art Deco-styled combination cigarette case with lighter. 4.5 inches tall. Made in St. Louis, Missouri. $50-$75

**MARATHON,** ca.1928. Alligator covered metal pocket lighter with a an unusual mechanism. When you turn the wheel, it turns another gear. The wick is in an unusual position. $80-$120

**MARATHON,** ca.1928. A two color enameled lift arm lighter. Made in the U.S. and 2.25 inches tall. $100-$120

**MAGIC & RELIABLE,** ca.1892. Two very early pocket lighters that used inventive means of creating a flame. These were known as pocket lamps. The Magic was made by the Magic Introduction Company or New York, using Koopman's patent of 1889. It used a spinning metal disk covered with fine crushed pyrite. An iron spring bar would press against it creating sparks to light a wick. The Reliable was made in Philadelphia and used a roll of fulminate of mercury caps to create a spark. Both are 3 inches tall. L-R $150-$200, $250-$350

**MARATHON,** ca.1935. Black enamel metal pocket lighter. Mechanism shows the long toothed ratchet that turns the striker wheel. $70-$90

**G.E. MARDINI,** ca.1927. Small pocket lighter. Note the protector on the snuffer arm. Made in France. $30-$40

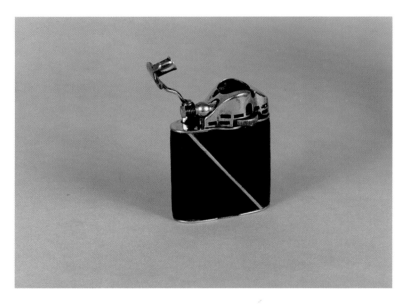

**MARATHON,** ca.1936. Leather covered metal pocket lighter with a highly unusual massive windguard. $80-$120

**MARS,** ca.1948. The "Trigger" model is a sleek modernistic, aluminum lighter. Made in the U.S. and 2 inches tall. $50-$75

**MATCHKING,** ca.1935. Striker table lighter with a clock. The striker pulls out from the top. 5 inches tall. Made in USA. $185-$240

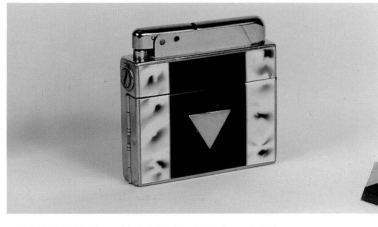

**MARATHON,** ca.1940. The "Ladylite" model cigarette case lighter. Made in the U.S. and 3.75 inches tall. $65-$95

**MAYFAIR,** ca.1930. An almost crudely designed lighter with a Bakelite body. 2 inches tall. $40-$65

**MAYFAIR,** ca.1924. A high quality man's pocket lighter marked "Delineator." $30-$45

**McMURDO,** ca.1946. Presentation table lighter in chrome. $75-$125

**NETOP**, ca.1928. An unusual small lift arm lighter. Made in the U.S. and 2 inches tall. $90-$110

**NOBLE**, ca.1955. Noble made this series of state and city lighters. One side shows a map or architectural monument, and the other shows the state bird and flower or other landmark. A collection of all 50 or more would be impressive. Made in Japan and 2 inches tall. $15-$30

**OMEGA**, ca.1948. A man's lighter with naked ladies on each side. The lighter is an inexpensive piece, but there is a lot of variety in these naked lady lighters that could make a nice collection. 2.25 inches tall. Made in Japan. $20-$25

**OLYMPIC**, ca.1932. Sort of an Olympic medal in Bakelite, it is marked, "Tenth Olympiad, California". 2.25 inches tall. $100-$150

**ORLIK LIGHTERS:** Founded in 1899, they began as a pipe maker. They moved on to making lighters about 1916, some of which were made in the U.S. and others in the U.K. Their lighters seem to disappear in the 1940s.

**ORLIK,** ca.1924. Man's pocket lighter. Made in England. $40-$50

**ORLIK,** ca.1926. Two variations of men's pocket lighters with windguards. They are slightly crude in appearance, but have a lot of personality. Made in England. $40-$50

**ORLIK,** ca.1924. Lady's pocket lighter with unusual ostrich leather covering. Made in England. $40-$50

**ORLIK,** ca.1934. Man's pocket lighter. The 1930s Orliks have kept the mechanism of the 1920s models, but the look is much smoother. Chrome-plated. Made in England. $40-$50

**PARK ROGERS,** ca.1931. A substantial pocket lighter. Push in the button for the semi-automatic action. 2.5 inches tall. $50-$75

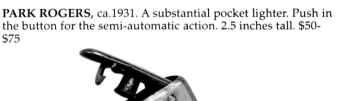

**ORLIK,** ca.1926. Enamel on gold filled metal pocket lighter. Lift arm mechanism. Back side is blue enamel. Made in England.

**PARKER,** ca.1928. This 14 inch tall sailboat is a large striker table lighter. The wand at the back of the boat is removed and scratched along the groove at the back of the boat. The sail appears to get in the way but can be moved to the side. $275-$325

**PARKER BEACON**, ca.1935. This handsome combination lighter and cigarette case in Bakelite is tough to refill with fluid, since the lighter is not removable. 4 inches wide. $150-$200

**PARKER BEACON**, ca.1940. An English-made pocket lighter in a Bakelite case. 2 inches tall. $50-$75

**POLAIRE**, ca.1933. An elegant table lighter 3.5 inches tall. Made in France. $50-$75

**POLO LIGHTERS:** 1930s to 1950s. English-made, everyday working man's lighters. They used a lift arm mechanism. At this time, they are relatively easy to find.

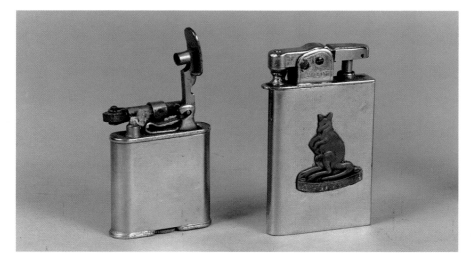

**POLO**, ca.1924 and 1948. The earlier Polo pocket lighters have more interesting lighter mechanisms than the later models. Made in England. $30-$40

**POLO**, ca.1938. Attractive silver-plated pocket lighter. English-made. $20-$30

**PP LTD.**, ca.1940. Sterling silver and gold pocket lighter. 2.25 inches tall. $75-$90

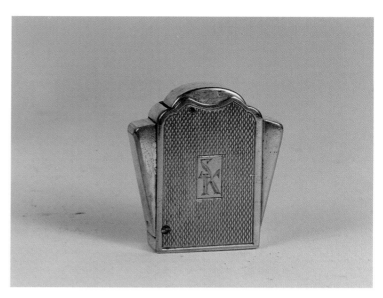

**QUERCIA**, ca.1937. An elegant lighter in the shape of a car grill. Push in the sides to light. 2.5 inches tall. Made in France. $150-$200

**PREMO**, ca.1933. This leather covered model lights in the middle. Made in the U.S. and 2.25 inches long. $65-$85

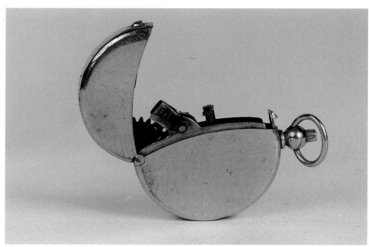

**R.K.**, ca.1919. An early automatic watch-shaped lighter. Made in Austria and 2 inches tall. $120-$175

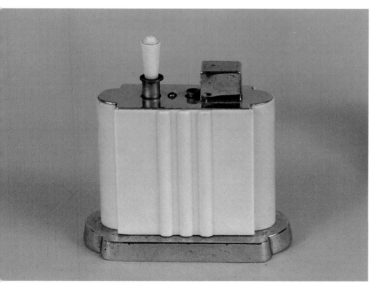

**PRESS-A-LITE**, ca.1930. A handsome table lighter that worked by removing the "wand" and pressing it down on the top button. This would activate the mechanism sending a shower of sparks at the wick on the end of the wand. About 4 inches tall. This was probably inspired by Ronson's Touch-Tip system. $125-$175

**R.K.**, ca.1930 & 1936. Two King lighters made in Austria with very different lighting mechanisms: push tab on the left (while turning plastic knob); and push button on the right.

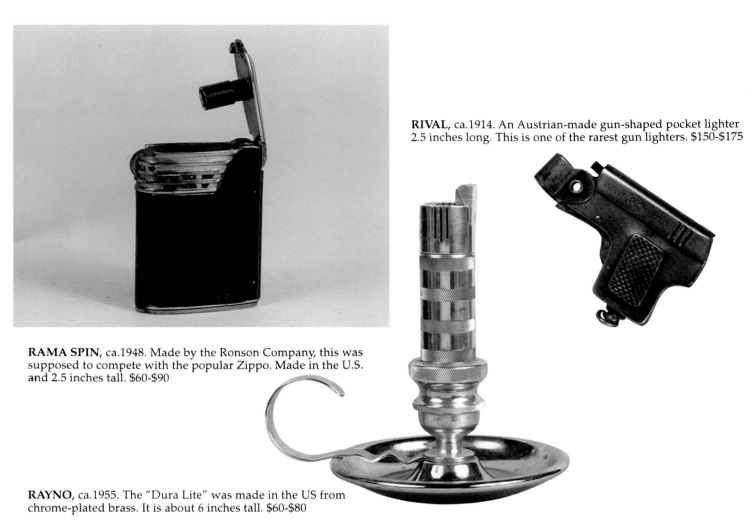

**RIVAL**, ca.1914. An Austrian-made gun-shaped pocket lighter 2.5 inches long. This is one of the rarest gun lighters. $150-$175

**RAMA SPIN**, ca.1948. Made by the Ronson Company, this was supposed to compete with the popular Zippo. Made in the U.S. and 2.5 inches tall. $60-$90

**RAYNO**, ca.1955. The "Dura Lite" was made in the US from chrome-plated brass. It is about 6 inches tall. $60-$80

**REGEL**, ca.1936. The Regeliter was a push button pocket lighter. This model was designed for a lady's purse and is about 1 inch tall. Made in the USA. $30-$40

**RONSON:** Started before the turn of the century by Louis V. Aronson (1870-1940) as the Art Metal Works, Newark, New Jersey. it used the name Ronson in the 1920s. The name was later changed to Ronson Art Metal Works, Inc. (1945) and then to Ronson Corporation (1954). Their first pocket lighter in 1913 was called the Wonderliter. It was a wand or striker type lighter. Wand lighters used a metal wand with a wick. The wand was dragged across the striker and the wick would light from the shower of sparks. Wands could be lost and replaced over time. They should fit snugly in their hole.

Ronson's first pencil lighter was manufactured in 1919. Looks like an Eversharp style pencil, but top contains a wand. There is a striker on the side of the pocket clip.

Their first automatic cigar lighter, the Banjo, was introduced in 1926-1927. Automatic means that one push would expose the wick, turn the striking wheel, and light the lighter. After the Banjo came the Standard (1928) and then the Princess (1929) which was smaller and was made for about 30 years. The early automatic lighters had exposed gears. The Standard and other early lighters were marked "Ronson De-Light". These used a screw that passed through the fulcrum from one side and screwed into the other side of the lighter mechanism. Later they used two screws, one entering from each side of the lighter to hold the mechanism.

The "Adonis" model was designed by Frederick Kaupmann and introduced in 1947 and was made for about 20 years. Their first butane models appeared in the early 1950s. Pocket models were called "Maximus" and table models had names starting with the letter "V".

**RONSON**, ca.1916. An early and very rare Ronson lion-shaped striker lighter made of cast iron. About 6 inches tall. $700-$900

**RONSON**, ca.1916. An early Ronson Bulldog striker lighter made of cast iron. About 6 inches tall. $200-$300

**RONSON,** ca.1919. The early "Wonderliter" lighter and original box. Made in the U.S. and 2.5 inches tall. $225-$275 w/o box, $750 w/box.

**RONSON,** ca.1920. Tipperary pup. This is an early Ronson striker lighter. The dog's tail is removed and scratched along a groove in the dog's back. $500-$600

**RONSON,** ca.1920. The Monkey bronze striker model. The area with the pipe is an ashtray. $175-$250

**RONSON,** ca.1920. "Pipe smoking Bulldog" model. An early striker model with the striker on the side. 5 inches tall. $175-$250

**RONSON,** ca.1920. An Art Metal Works inkwell and pen holder. 6 inches wide. $60-$80

**RONSON,** ca.1920. An unusual Aronson Art Metal Works marked Ronson desk pipe holder. It is unusual that the central post holds a box of matches rather than a lighter. 12 inches across. $100-$125

**RONSON,** ca.1923. A Ronson incense burner that is missing its cover. 8 inches tall. $80-$100

**RONSON**, ca.1929. A "De-Light" pocket model in gold-plated brass. This round shape was very popular. $30-$40

**RONSON**, ca.1927. The first Ronson automatic lighter called the Banjo. A mint example in its box. Note the Art Metal Works logo on the box. $450-$550 w/o box.

**RONSON**, ca.1928. "Bulldog Ashtray" model. An nice striker model with the striker on the front. 5 inches tall. $175-$250

**RONSON**, ca.1927. Table Banjo lighter 4.5 inches tall. Leather on silver-plated metal. $350-$450

**RONSON,** ca.1928. An early De-Light Tabourette table lighter in silver-plated metal. 4 inches tall. $125-$175

**RONSON,** ca.1928. This Ronson tank striker table lighter is only about 3 inches tall. Made in the U.S. $350-$450

**RONSON,** ca.1929. The "New Yorker" model De-Light table lighter in an Art Deco Bakelite material. 3 inches tall. $125-$175

**RONSON,** ca.1929. Table De-Light lighter 5.5 inches tall. Black enamel and chrome.

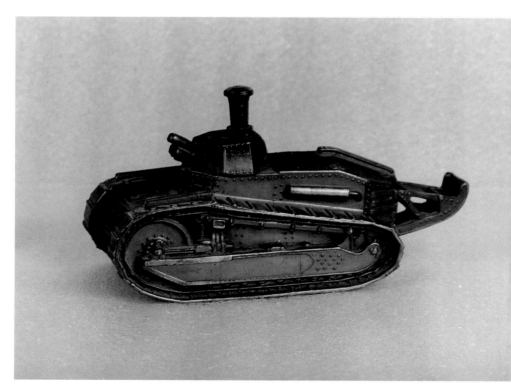

**RONSON,** ca.1930. Tank model striker lighter made of a bronzed white metal. An unmarked Ronson model. 6 inches long. $450-$550

**RONSON,** ca.1929. The Ronson "De-Light" with a chrome overlay. Made in USA. $350-$425

**RONSON**, ca.1934. One of the most sought after Ronson lighters: The Bartender Touch-tip. In the Touch-tip system, a permanent match (called the 'Wand") is withdrawn and pushed down on a mechanism that causes a shower of sparks to fall on the wick end of the wand. It also holds cigarettes in the two side compartments which flip back. About 6 inches tall. $1300-$1600

**RONSON,** ca.1934. Combination cigarette lighter, cigarette case and lady's compact. A problem with including a lady's compact in the case is that the dusting powder got into the cigarettes and the lighter fluid often tainted the powder. 5 inches tall. $100-$150

**RONSON,** ca.1934. The Lincoln striker model. The original sculpture was done by Gutzon Borglum in 1910 and sits in front of the Essex County, New Jersey courthouse. $175-$225

**RONSON.** Three Ronson lighter accessories. Two are flint holders and one is a service kit. $5-$10

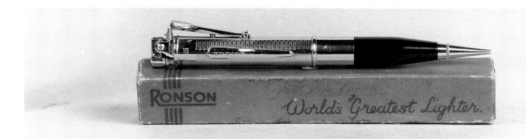

**RONSON,** ca.1934. The Ronson pencil lighter was made in enough quantity so that you should have no trouble finding one. It was well-made and perfect for the person who wrote and smoked. This is the first automatic model in the rarer metal & plastic. All metal is more common. During the 1930s many items were combined into one piece. There were pen/pencil combinations, compact/lighter and cigarette case/lighter combos. Initially, they were popular, but the public soon lost interest. They found the combos to be heavier or bulkier than the individual item. $40-$60

**RONSON,** ca.1935. A black & chrome Touch-Tip model with a clock. $400-$450

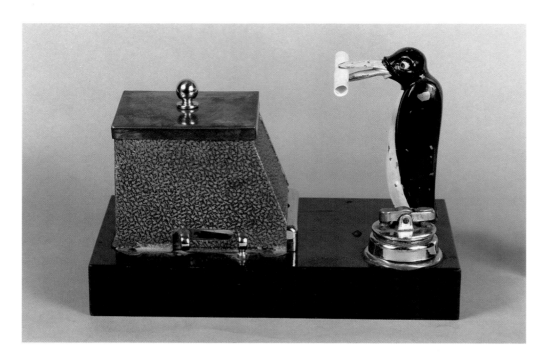

**RONSON**, ca.1935. The Penguin "Pic-a-Cig" lighter. The handles were squeezed to get the penguin to lean over and pick up a cigarette. 6 inches tall. $500-$700

**RONSON**, ca.1935. Touch-Tip table lighter. 3.5 inches tall. $75-$125

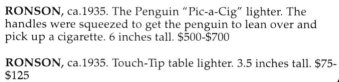

**RONSON**, ca.1935. A difficult to find Touch-Tip table lighter called "The Barmaid". This was made to compliment the Bartender Touch-Tip. $1300-$1600

**RONSON**, ca.1935. The "Octette" Touch-Tip model lighter in a rare combination with a silver-plated ash tray. 9 inches wide. $300-$350

**RONSON**, ca.1935. Pencil lighter that used a striker mechanism that fit into the eraser end of the pencil. 6 inches long. Made in USA. $225-$275

**RONSON**, ca.1935. The "Gem" lighter in sterling silver. Made in the U.S. and 2.25 inches tall. $175-$225

**RONSON**, ca.1936. The Ronson "Mastercase" Deco enameled combination lighter/cigarette case. Made in USA. $60-$85

**RONSON**, ca.1936. The Ronson "Magnapact" Deco enameled lighter, cigarette case and built-in compact was only made for a few years. The material was called "Dureum" which was a gold-plated metal. Made in USA. $350-$450

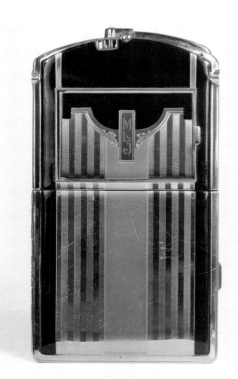

**RONSON**, ca.1936. The Ronson "Smartset" Deco enameled combination lighter/cigarette case. Made in USA. $120-$150

**RONSON**, ca.1936. The Ronson "Beauticase" Deco enameled combination lighter/cigarette case shown with the lighter in closed position and open position. Made in USA. $300-$350

**RONSON**, ca.1936. The Ronson "Heart", a Deco enameled lighter made of Dureum. Made in USA. $450-$550

**RONSON**, ca.1936. A very handsome Touch-Tip table lighter from Ronson that sported a desk clock with decorative Scotty dog. To wind the clock, you pulled a string on the top. 6 inches tall. $800-$950

**RONSON**, ca.1936. A Touch-Tip "Deluxe" black & chrome table lighter. 4.5 inches tall. $200-$250

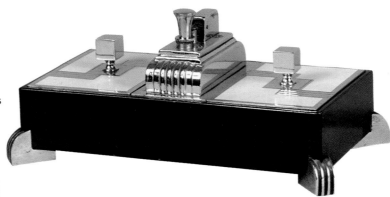

**RONSON**, ca.1936. A double cigarette box Touch-Tip table lighter with compartments for cigarettes. 8 inches wide. $250-$350

**RONSON**, ca.1936. An unusual Touch-Tip table lighter with an 8 day desk clock. 6.5 inches wide. $600-$800

**RONSON**, ca.1936. The Ronson "Masterpact" model was a compact, cigarette case and lighter combination available in different finishes. Made in the U.S. and 4.75 inches tall. $175-$225 each

**RONSON.** A great group of miscellaneous Ronson lighters made from the 1930s to the 1950s.

**SCRIPTO:** 1950s to 1960s. This Atlanta, Georgia company made view lighters (see through bodies). Most are in the $15 to $25 range with very few going to $100. They are becoming more popular with collectors, as they have great variety and their price range is reasonable. As with Zippo, salesman's samples and the advertising models are more desirable.

**SMOKERETTE**, ca.1930. An unusual toaster style automatic cigarette dispensing table top lighter. Plugged in, the case was filled with cigarettes and would dispense them lit. 6 inches tall. $75-$100

**SCRIPTO**, ca.1955. A salesman's sample case showing the colors in which Scriptos were available and different advertising concepts that a customer could order. $200-$250

**SPEED**, ca.1940 & 1950s. An eastern Indian motif model and a chrome-plated model. A spur on the half moon-shaped top is the trigger to light these interesting lighters. 2.5 inches tall. $50-$70

**SCRIPTO**, ca.1957. The Scripto used a visible fuel area that could contain advertising or a fishing fly. This example contains rare advertising for Scripto. Made in USA. $45-$65

**SS & CO.**, ca.1930. A water tank model striker lighter with ashtray made in England. 6 inches tall. $125-$175

**TASSEL-LITER**, ca.1935. A very strange concept. The bulb holder would screw into a light fixture, ideally a fixture over a reading/smoking chair. When you wanted to light your smoke, you grabbed the tassel and pushed a button on the side. Your smoke was lit on the glowing element. $75-$100

**STESCO**, ca.1940. A pretty good quality lighter, but the calendar makes this piece unusual. 2 inches tall. $30-$40

**TEAM**, ca.1932. An Austrian table lighter with Deco styling. 3 inches tall. $40-$60

**SWANK**, ca.1960. A Japanese made combination cigarette lighter/slide viewer. 4 inches wide. $25-$30

**TAN**, ca.1938. A sturdy man-size pocket lighter in ribbed silver plate. Made in Switzerland and 2.5 inches tall. $100-$175

**THORENS:** 1920s to 1960s. These Swiss-made lighters of the highest quality are almost in the class of Dunhills. Thorens was a music box maker that started in 1883. Their most sought after models are of the 1920s to 1930s era. The company moved into making record turntables, speakers and other music items.

**THORENS,** ca.1930. A sterling silver bell-shaped table lighter with a semi-automatic action. $125-$140

**THORENS,** ca.1932. A great Art Deco bronze nude holding a semi-automatic table lighter. Made in Switzerland and 3.75 inches tall. $250-$300

**THORENS,** ca.1930. A chrome-plated table lighter with a semi-automatic action. $40-$60

**THORENS,** ca.1939. Another great Art Deco style table lighter with a flying crane. Made in Switzerland and 4.25 inches tall. $100-$125

**THORENS,** ca.1940. A nicely designed circle lighter in chrome. This is a rare lighter with an unusual mechanism and shape. 2 inches tall and made in England. $135-$160

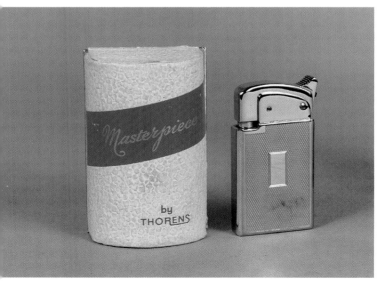

**THORENS,** ca.1949. "Masterpiece" model pocket lighter. $75-$100

**THORENS,** ca.1953. A chrome-plated automatic lighter called the "Vendette". Push the side button to light the lighter. 3 inches tall. $40-$60

**1000 ZUNDERS,** ca.1948. A high quality sterling silver lighter with the Mercedes Benz logo. German-made and 2 inches tall. $150-$175

**TRIANGLE:** 1928 to about 1940. A high quality lighter that is not too easy to find. They are known for their nice watch/lighter models. The bottom is usually shaped as an elongated hexagon.

**TRIANGLE,** ca.1925. An American-made chrome-plated pocket lift arm lighter with a Masonic insignia. $40-$60

**TRIANGLE,** ca.1928. A nice example of the Triangle lift arm, leather covered lighter/watch combination. 2 inches tall. $175-$225

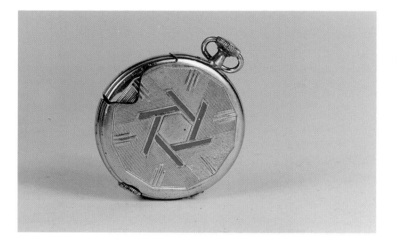

**TIKTAK,** ca.1936. An enameled watch-shaped lighter. Push in the stem to light the lighter. Made in Austria and 2 inches tall. $125-$175

**TRIANGLE,** ca.1934. A pocket lighter with a built in watch in silver-plated metal. 2 inches tall. $125-$150

**TIMELITE,** ca.1950. This would probably be for home use. Picking up the hand piece and pushing the button would heat the element for lighting a cigarette. Woe to the drunk trying to use this phone to call home. $100-$125

**UNKNOWN:** There are thousands of variations of lighters by unknown makers. Here are a few notes on those lighters:

French lighters made in 1911 were affixed with a copper "tax stamp" tag. The tag is dated 1911 and may have been in use for several years thereafter. During the 1920s and 1930s, French lighters were affixed with a silver "tax stamp" tag usually saying "Ministre Finance". Some were also stamped "BL" indicating a luxury tax.

Belgium lighters were required to have a metal tax stamp from 1923 into the 1930s.

The Japanese lighter industry grew dramatically in the 1950s and 1960s. Several American companies took advantage of the cheap labor in Japan and set up cigarette lighter factories. Japanese lighters made between 1945 and 1952 were stamped "Made in Occupied Japan". After the occupation, Japanese lighter manufacturers began producing thousands of models of lighters. By the end of the 1960s, the quality was among the best that could be found worldwide. Japan led the world in lighter innovation in the 1970s. With the decline in smoking, manufacturers have cut back spending the money on innovation.

**Unknown,** ca.1900. Devil's head table lighter. These were probably used in public places such as a bar or restaurant. The wick would be lit and allowed to burn. Each horn could be pulled out. Attached to a wire at the base of the horn was another wick that would be lit from the main flame. This was used to light the cigar or pipe. $90-$110

**Unknown,** ca.1910. An early automatic pocket lighter made in Germany of sterling silver. The release of the lid turns the sparking wheel.

**Unknown,** ca.1890. A silver-plated cigar lighter. The case was filled with alcohol and the central wick was lit. The 4 individual rods would be lit from the central flame and then used to light a cigar or pipe. 5 inches tall. $100-$150

**Unknown,** ca.1890. A bronze cigar lighter. The case was filled with alcohol and the central wick, between the ears, was lit. The mouth rod would be lit from the central flame and then used to light a cigar or pipe. Unusual that the tail is grooved for use in lighting a match. 6 inches tall. $100-$150

**Unknown,** ca.1915. Probably used in a public place such as a bar or restaurant. The wick would be lit and allowed to burn. The side wands could be pulled out. Attached to a wire at the base of the wand was another wick that would be lit from the main flame. Used to light the cigar or pipe. It is about 8 inches tall and was made in England. $250-$350

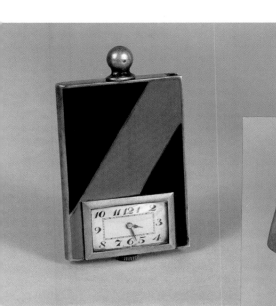

**Unknown,** ca.1916. A very rare early enamel-on-silver French-made striker lighter with an inset watch.

**Unknown,** ca.1916. An early push button pocket lighter made in brass. Made in Austria.

**Unknown,** ca.1917. A very unusual ring lighter made in brass with a sliding cover plate. Probably made in Austria.

**Unknown,** ca.1916. An early Austrian-made lighter in the shape of a Bulldog about 4 inches tall. When this metal bodied, glass eyed, dog's tail is pushed down, the head pops open lighting the wick. A particularly rare model.

**Unknown,** ca.1918. This complex piece of trench art supports the theory that much of the trench art was not made by soldiers in the trenches. 8 inches tall. $150-$175

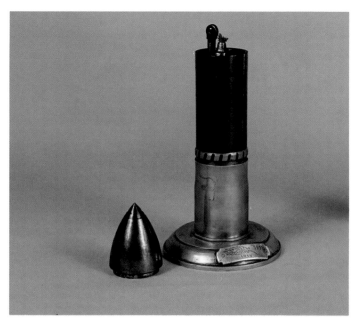

**Unknown,** ca.1918. Trench art lighter using a shell and projectile as the lighter holder. It is 8 inches tall. $90-$110

**Unknown,** ca.1918. A brass trench lighter covered with coins. The coins include an 1898 Indian head penny and a 1917 Lincoln cent. 2 inches tall. $75-$125

**Unknown,** ca.1918. Two trench lighters. The left one is American-made and the right one was most likely made in France during World War I. These are typical of this popular style. Patriotic motifs adorn both sides. The one on the right is made from a large nut. $60-$90

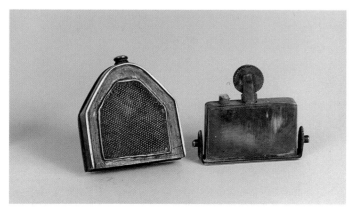

**Unknown,** ca.1918. An unusual piece of trench art made in the shape of a truck grill. Probably made in USA. $300-$350

**Unknown,** ca.1919. Trench lighter probably French-made during World War I. Typical round style. Patriotic motifs on one side and an airplane motif on the other. About 2 inches round. $75-$100

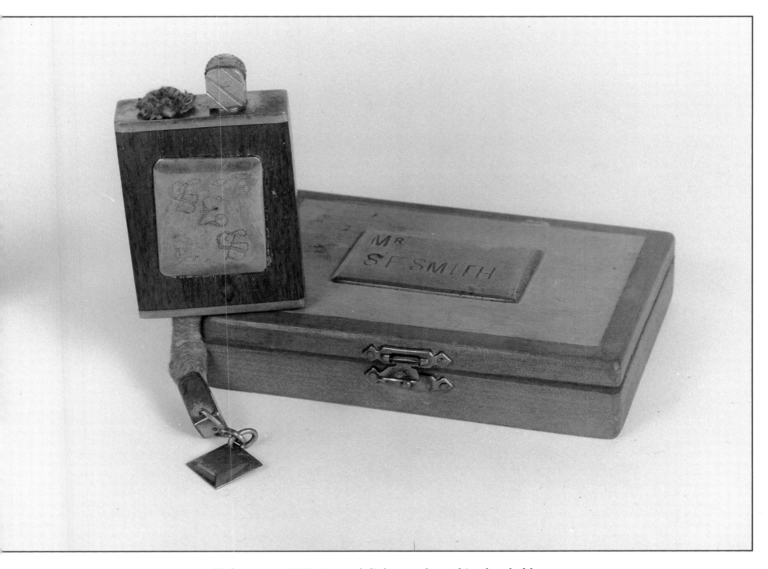

**Unknown,** ca.1918. A trench lighter and matching box holder. The style is the early rope lighter. The trick of using one of these was to get the rope smoldering, light the cigarette or cigar and then snuff out the smolder with the metal end of the rope. The lighter is 3 inches tall without the rope. $175-$225

**Unknown,** ca.1919. An unusual pen-shaped trench lighter nicely made of brass. 5 inches tall. $75-$100

**Unknown,** ca.1919. Trench art table lighter. The idea behind trench art pieces is that they were made by soldiers in the trenches using artillery shell brass. More realistically, most were made by local craftsman who salvaged brass shell casings and made souvenirs for the soldiers. $75-$100

**Unknown,** ca.1919. Brass lighter in the candlelight style. 6 inches tall. $40-$60

**Unknown,** ca.1919. There is a great image on this trench lighter: the Spanish Conquistador. 2 inches. $60-$90

**Unknown,** ca.1919. Trench lighters, possibly French, made during World War I. Variations on the round style. About 2 inches round. $60-$90 each.

**Unknown,** ca. 1934 & 1919. Two typical trench-style lighters. The one on the left has a George the VI coin. Since George the VI was not king until 1934, the piece is much later than the World War I vintage that it appears to be. The shell style is 4 inches tall. $60-$90 each.

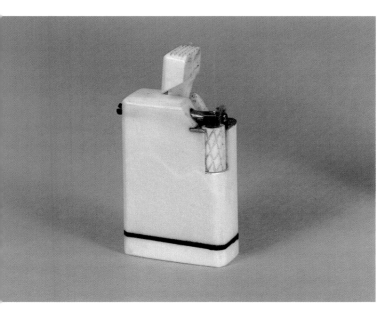

**Unknown,** ca.1929. Table lighter in the shape of Charles Lindbergh's plane, with an 5.5 inch wingspan. It lights by rolling the plane. A spur on the wheel hits a trigger that pops the lighter mechanism. Very rare with rubber wheels. Made in Germany.

**Unknown,** ca.1929. A very unusual all ivory pocket lighter made in England. $80-$110

**Unknown,** ca.1929. An exceptional silver-plated lift arm lighter with enamel paintings of airplanes on both sides. One plane is the Lindbergh type and the other is a clipper ship style. It may have been made to mark Lindbergh's flight. Made in Austria and 2 inches tall. $300-$350

**Unknown**, ca.1930. A store or bar lighter. The nude with a ball or at water's edge were popular motifs that were considered art rather than solely a "man's lighter". The electric power indicates that this was for heavy duty use rather than something that would be used in a home. A button on the back when pushed, would cause the mica covered element to heat up to light a cigarette or cigar. $50-$80

**Unknown**, ca.1930. Genie lamp striker style table lighter about 6 inches wide. The lamp's handle is removed and scratched along the groove on the base. The flame cap can be removed. Underneath it is a wick that can be lit to light other cigarettes or used as a table lamp. Made in Austria or Germany.

**Unknown,** ca.1930. A very attractive nude, figural, gold-washed metal, striker lighter. 8 inches tall. $250-$300

**Unknown,** ca.1930. A very strange, but nicely made, torch lighter. There are four wicks in the end. The back end is pushed forward creating a shower of sparks. 12 inches long when closed. $50-$75

**Unknown,** ca.1930. Many stores would offer a free fill of lighter fluid. These hard to find lighter fluid pumps came in many shapes and sizes. This especially handsome one is in the shape of a gas pump and graced the counter of Kuppenheimer's Good Clothes store. 12 inches tall. $300-$400

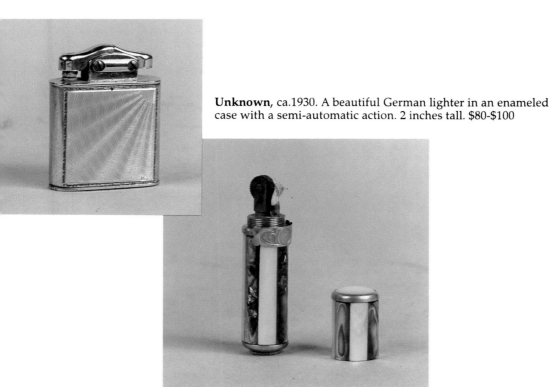

**Unknown**, ca.1930. A beautiful German lighter in an enameled case with a semi-automatic action. 2 inches tall. $80-$100

**Unknown**, ca.1930. A French lipstick style lighter with the French tax stamp on the front. Covered in mother-of-pearl and abalone and 2.5 inches tall. $35-$50

**Unknown**, ca.1932. Metal camel table lighter. There are quite a few camel table lighters, possibly to light Camel cigarettes. 6 inches long. $30-$40

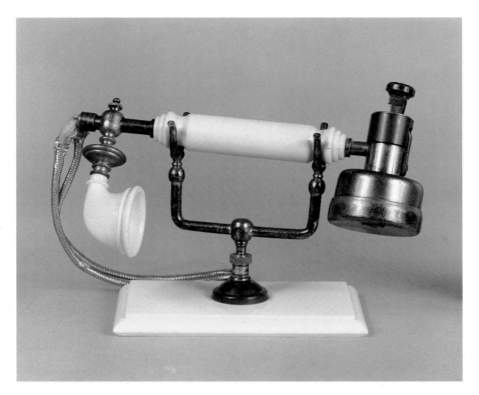

**Unknown,** ca.1934. An unusual telephone lighter. The lighter is in the earpiece assembly. Possibly French-made. $30-$40

**Unknown,** ca.1934. Celluloid covered table lighter with a hotel's name on the side. 4.5 inches tall. Made in Germany. $40-$50

**Unknown,** ca.1934. Mother-of-pearl pocket lighter. 2.5 inches tall. Made in Japan. $20-$25

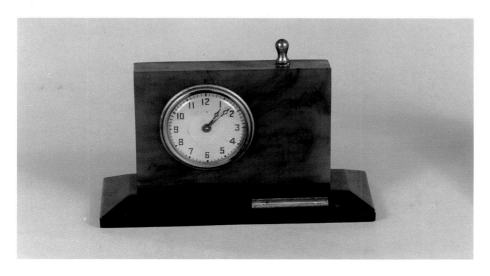

**Unknown**, ca.1935. A desk clock with a striker lighter. 5 inches tall. This piece was made in America.

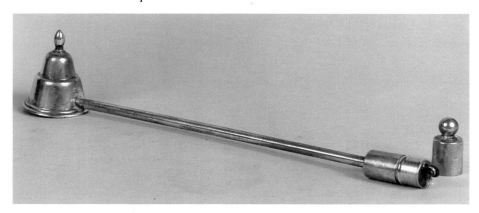

**Unknown**, ca.1935. The elegant sterling silver candle snuffer has a lighter in the end. 12 inches long. $40-$45

**Unknown**, ca.1935. A super figural caddie and golf bag lighter. Actively collected by golf collectors. $200-$225

**Unknown**, ca.1935. French decorative striker lighter, with a Champagne bottle and two cups on a tray. The top of the bottle pulls out to strike on the front. It is 4.5 inches tall. $175-$200

**Unknown**, ca.1935. Striker lighter made in the shape of a gas pump with the "Flying A" insignia. It is 4.5 inches tall and was made in Germany. $175-$200

**Unknown**, ca.1935. This pocket lighter was made in England and is interesting for its long turning wheel. $20-$25

**Unknown**, ca.1935. This sterling silver with gold wash ensemble (lighter & cigarette case) was made in Germany or Austria. The quality and style is very similar to the Dunhill's of the period. The box is marked with a Spanish jeweler's name. 4 inches long. $600-$700

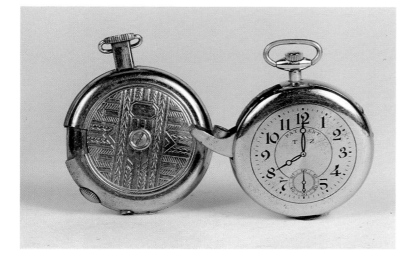

**Unknown**, ca.1935. Two chrome-plated watch-shaped lighters. Push in the stem to light the lighter. 2 inches tall. $65-$85

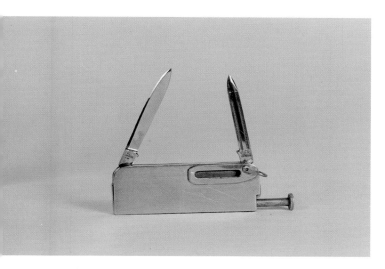

**Unknown**, ca.1939. An alarm clock-shaped lighter. 2.5 inches tall. Made in Japan. $150-$175

**Unknown**, ca.1938. A gentleman's pocket accessory. It contains a nail file, a knife and a striker lighter. 1.25 inches tall. $65-$85

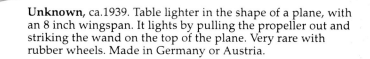

**Unknown**, ca.1939. Table lighter in the shape of a plane, with an 8 inch wingspan. It lights by pulling the propeller out and striking the wand on the top of the plane. Very rare with rubber wheels. Made in Germany or Austria.

**Unknown**, ca.1940. Metal figural striker lighter in the shape of a jockey and horse, about 9 inches tall. $125-$160

127

**Unknown**, ca.1944. This Egyptian-made lighter used an odd grooved ball to spark the flint. Made of machined aluminum, it may have been made by a machine shop serving the U.S. army during World War II. It is about 4 inches wide. $75-$110

**Unknown**, ca.1944. This British-made lighter was made of machined aluminum. It is the World War II equivalent of the World War I trench lighter. Probably made out of aircraft aluminum by a machine shop during the War. It is typical of the lighters made in Egypt and is about 4 inches tall. $60-$80

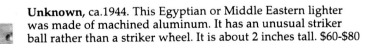

**Unknown**, ca.1944. This Egyptian or Middle Eastern lighter was made of machined aluminum. It has an unusual striker ball rather than a striker wheel. It is about 2 inches tall. $60-$80

**Unknown**, ca.1946. A French table lighter in a bamboo case. 3.75 inches tall. $35-$50

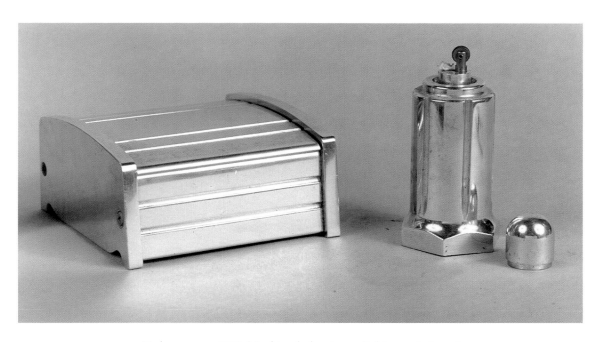

**Unknown,** ca.1945. Machined aluminum lighter and cigarette case. American-made. It is about 4.5 inches tall. $40-$60

**Unknown,** ca.1947. "Call for Phillip Morris" bellhop about 4 inches tall. Made in Occupied Japan. This lights automatically when lifted from the table. $100-$1250

**Unknown**, ca.1947. A lighter in the shape of a fan about 5 inches tall. This piece was made in Occupied Japan. Turn blade to light.

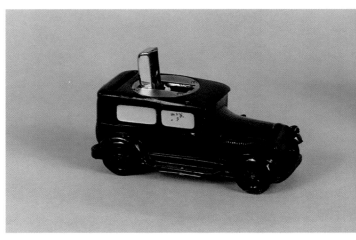

**Unknown**, ca.1948. Taxi lighter about 5 inches long. Made in Germany. This is another very rare lighter.

**Unknown**, ca.1947. A machined aluminum with plastic and coin inserts lighter, ashtray and cigarette holder. Made in North Africa.

**Unknown**, ca.1947. A small typewriter lighter. It lights when the space bar is pressed. 1.5 inches tall and made in Japan. $150-$200

**Unknown**, ca.1947. A small baseball lighter. 2.5 inches tall and made in Japan. $75-$100

**Unknown**, ca.1948. Metal Peeing Boy lighter. This belongs to the group of lighters that were popular in men's clubs and college fraternities. A striking wheel on the lad's rear shoots a spark forward lighting a wick covered by his hand. Probably not used to light a lady's cigarette in polite society. About 3 inches tall. $90-$100

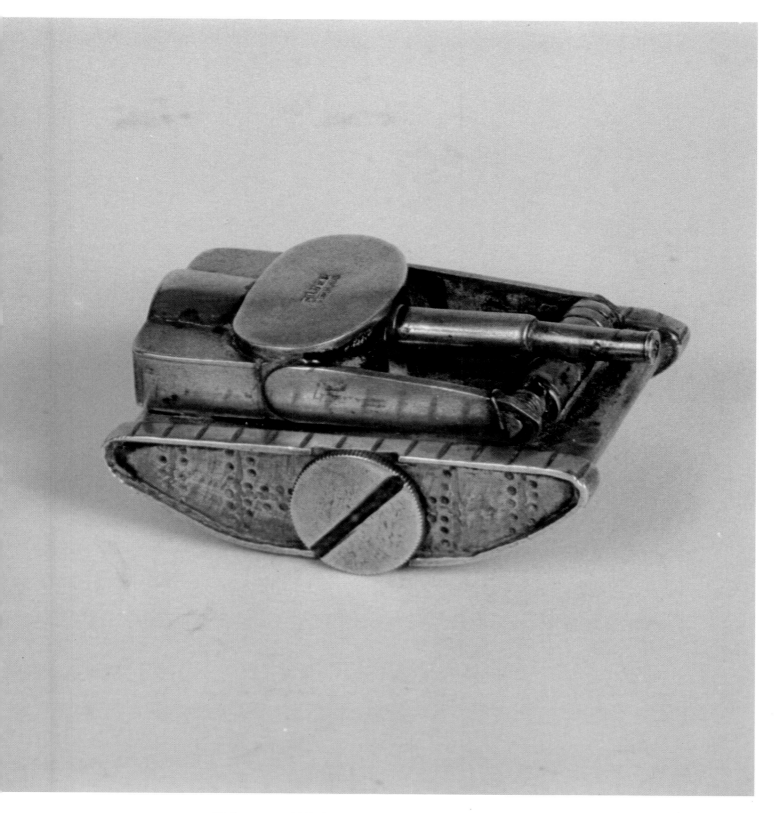

**Unknown,** ca.1948. An exceptionally nice, small Mexican-made silver lighter in the shape of a tank. The gun contains the spring for the flint and the mechanism is under the rear top plate. 2 inches long. $650-$700

**Unknown**, ca.1948. An enameled aluminum lighter with an unusual feature to aid in pipe lighting. The extra piece would unscrew, be dipped into the flame and then used to light a pipe. 2.75 inches tall. $75-$100

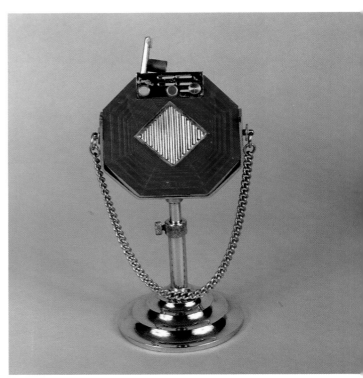

**Unknown**, ca.1949. Plastic and metal lighter in the shape of a microphone about 6 inches tall. Made in Occupied Japan. $150-$200

**Unknown**, ca.1949. Two very well-made chrome metal lighters in the shape of American cars about 5 inches long. Made in Occupied Japan. The one of the right is a friction powered toy. $200-$250 each.

**Unknown,** ca.1949. Plastic and metal lighter in the shape of record player about 4 inches long. Made in Occupied Japan. The knob is turned to light. $200-$250

**Unknown,** ca.1949. Plastic and metal lighter in the shape of console radio about 4 inches tall. Made in Occupied Japan. $200-$250

**Unknown,** ca.1949. "Silent Flame" type leather covered table lighter with cigarette dispenser. As the top is turned, the cigarettes, which are spring loaded, pop up through the hole. $40-$60

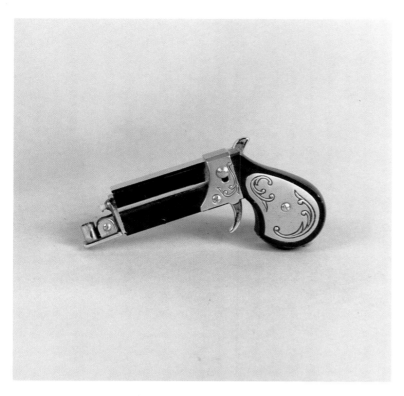

**Unknown,** ca.1950. Metal lighter in the shape of a derringer. 3 inches long. Made in Japan. $20-$25

**Unknown,** ca.1950. Ceramic and metal lighter in the shape of a cat and lamp. 7 inches long. Made in Japan. $10-$15

**Unknown,** ca.1950. A cowboy hat semi-automatic lighter. 4 inches long. Made in occupied Japan. $30-$45

**Unknown,** ca.1950. Julie holds the ultimate in large table top lighters. It was probably used in a cigar store window as a display piece, although all the parts are right to allow it to work. $30-$40

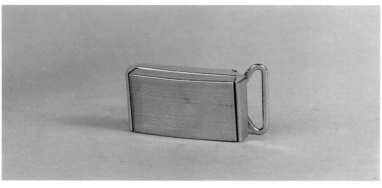
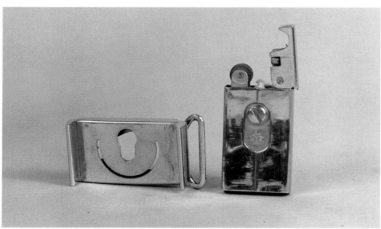

**Unknown,** ca.1950. An very unusual belt buckle lighter. 2.5 inches long. Made in occupied Japan. $40-$50

**Unknown,** ca.1950. A wonderful bellows lighter. When the bellows is compressed, the lighter lights. 6 inches long. Made in Japan. $10-$20

**Unknown,** ca.1952. An aluminum lighter with a watch. 3 inches tall. $100-$150

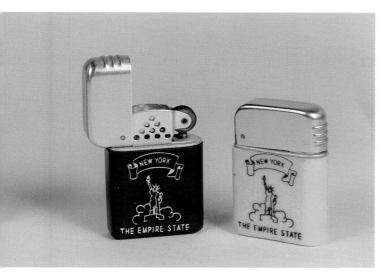

**Unknown,** ca.1958. These interesting inexpensive lighters sport Zippo-type mechanisms. Probably sold in the tourist shops on Broadway. 2.5 inches tall. $15-$25

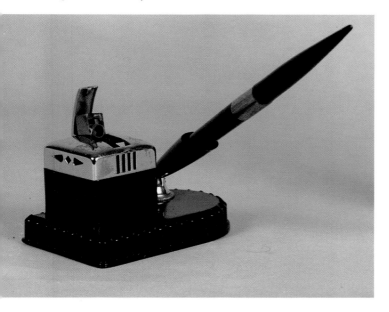

**Unknown,** ca.1960. Desk pen holder and lighter made in Japan. $40-$60

**Unknown,** ca.1960. Plastic and metal lighter in the shape of stock ticker about 6 inches tall. Made in Japan. $35-$50

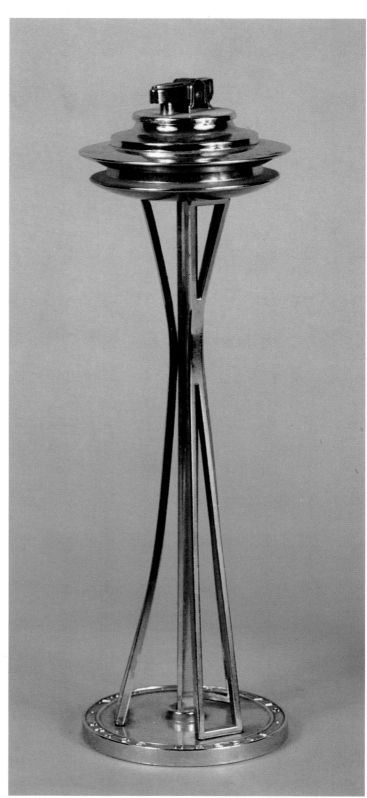

**Unknown,** ca.1960. Table lighter in the shape of the space needle erected for the Seattle, Washington World's Fair. Made in Japan. 10 inches tall. $35-$50

**Unknown**, ca.1960. Plastic and metal lighter in the shape of a car. When rolled forward, the cigarette drawer opens. 7 inches long. Made in Japan. $20-$30

**Unknown,** ca.1960. Table lighter made in the shape of a suit of armor. 10 inches tall and made in Japan. $20-$30

**Unknown,** Believed to be a Feu-Follet, ca.1960. This large (7 inches tall and 4 inches wide) candle is secretly a butane lighter that works when picked up. A battery operates the spark generator. Made in France. $175-$225

**Unknown,** ca.1960. An oil derrick lighter. 6 inches tall. Probably made in Japan by Shields. $20-$30

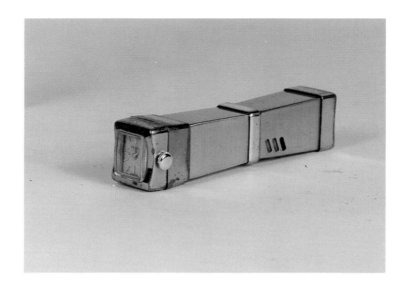

**Unknown,** ca.1960. A watch/lighter combination. The two ends are pressed in to light the center section. 3 inches long. $40-$65

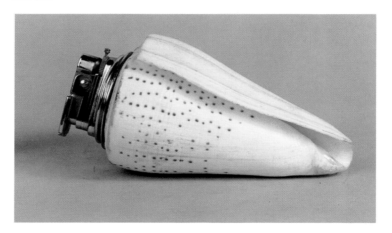

**Unknown,** ca.1970. A "welk" butane lighter. 4 inches long. An entire collection of lighters in seashells would be a bizarre collection. Made in Japan. $10-$15

**Unknown,** ca.1979. An unusual Nikon camera novelty lighter. 3.5 inches tall. Made in Japan. $15-$20

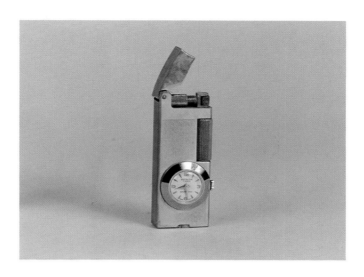

**VD,** ca.1948. A later lighter/watch combination. Made in Switzerland and 3 inches tall. $90-$135

**VAN LANSING CO.,** ca.1935. A hard to find lighter fluid pump. It dispensed fluid into your lighter for one cent. It was called the Van-Lite and the company was located in Pittsburgh, PA. 19 inches tall. $200-$300

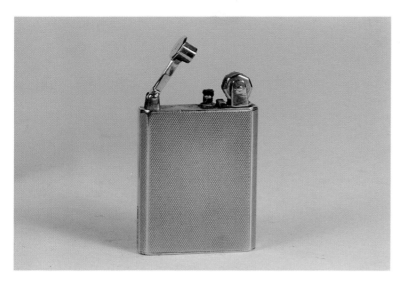

**VICEROY,** ca.1928. Lighter in 9 kt gold. Made in the England and 2.5 inches tall. $950-$1100

**WOND-O-LITE,** ca.1947. An unusual lighter in the shape of a twin lens camera about 4 inches tall. Made in Occupied Japan.

**WEBSTER,** ca.1929. Two views of the lighter in closed position and with the windscreen and cover lifted. Made in sterling silver in the USA. $600-$700

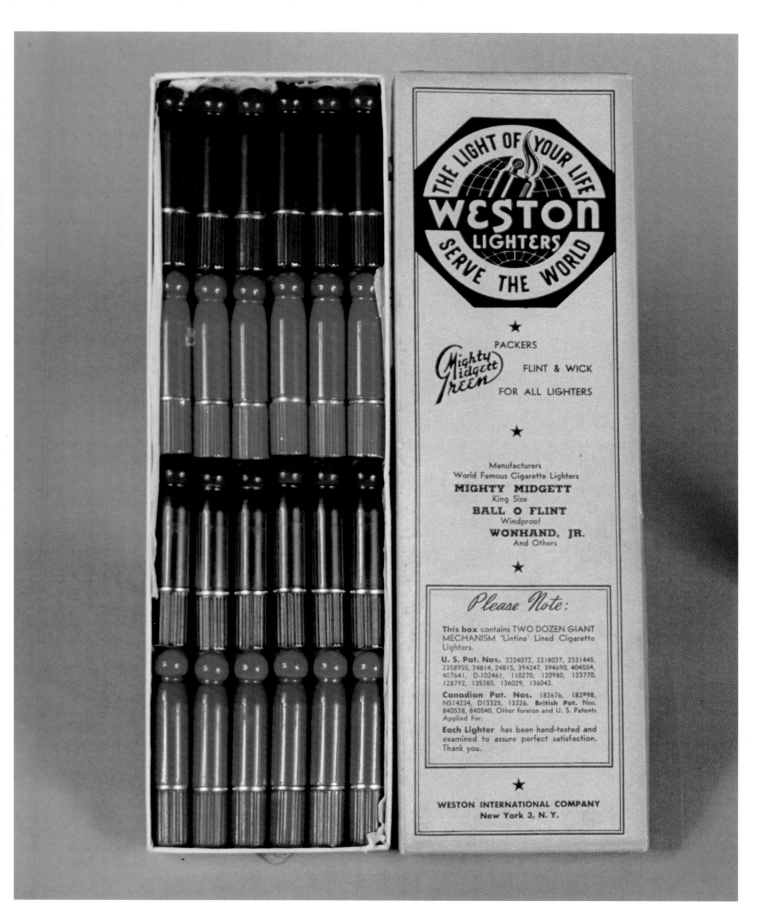

**WESTON**, ca.1950. A full box of colorful Weston lighters. The box indicates that they also made the Mighty Midgett, Ball O Flint and Wonhand, Jr. lighters. $200-$250 for all.

**WRIGHT,** ca.1911. An early hand-chased and engraved automatic pocket lighter made of highly engraved Sterling silver. $200-$250

**ZIPPO:** 1932 to present. Collectors of Zippo lighters have become one of the largest group of lighter collectors. The beauty of the Zippo may elude some collectors since they almost all come in the same style of case and use the same mechanism, one which has few parts to intrigue those who love complex mechanisms. It seems that Zippo lighter collectors are collecting something different than other lighter collectors. They collect social history.

Zippos advertised or showed products or things, many that only existed for a short period of time. They reflect what happened during the years since the 1930s. There are lighters that were made to honor battles, aviation flights, space flights, award winners, and everyday events. They were used by soldiers in all the wars since the Second World War. They were available in custom designs in lots as small as 50 pieces so that any group of people could memorialize their achievements. They were reliable and came with the best guarantee: If it breaks, Zippo will fix it for free. Like small clay tablets of ancient Egypt, they will tell social history thousands of years into the future.

**ZIPPO,** 1935. Zippo company invoice.

**ZIPPO,** George G. Blaisdell, founder of Zippo.

**ZIPPO,** 1936. A very early Zippo "Boy fishing from the dock" lighter. This is the only known example of this model. $1900-$2100

**Dating Zippo Lighters:**

**1932:** First model created by George Blaisdell. It is about one quarter inch longer than later models.

**1933 to 1935:** The hinge connecting the top and bottom of the lighter case is soldered onto the outside of the case. The hinge is made up of 3 barrels, one center, usually connected to the bottom hinge plate and 2 outside, connected to the top hinge plate. The bottom of the lighter case is flat and the edges are squared off. The windscreen has 16 holes and the cap pressure bar pivot pin area is a part of the windscreen.

**1936:** The hinge is still soldered onto the outside of the case and is made up of 4 barrels. The bottom of the lighter case is still flat and the edges are squared off.

**Mid to late 1936 to 1943:** The hinge is now soldered on the inside of the case and is made up of 4 barrels. The bottom of the lighter case is still flat and the edges are squared off or rounded. 1937 is the beginning of the brass drawn case with a more rounded top and bottom. Formerly, the top was flat and soldered into place.

**1943-1945:** The hinge is made up of 3 barrels. The bottom of the lighter case is slightly rounded and the edges are rounded.

**1946-1950:** The bottom of the lighter case is changed. It has a concave framed look, that is, there is an indented area where the imprint is located and the edges are rounded. In 1946, the windscreen is made with 14 holes. In 1947, the windscreen is made with 16 holes and the metal supporting the wheel now connects to the top of the windscreen. The words ZIPPOMFG, on the inside unit have no space between ZIPPO and MFG.

**1950 forward:** The patent number is changed to 2517191 from 2032695.

**1951 forward:** The hinge is made with 5 barrels.

**1957 forward:** There is now a code for year of manufacture on the bottom of the case as follows:

ZIPPO, 2 pages showing the dating chart of Zippos.

### ZIPPO LIGHTER IDENTIFICATION CODES

| YEAR | REGULAR LEFT | REGULAR RIGHT | SLIM LEFT | SLIM RIGHT |
|---|---|---|---|---|
| 1932 | Patent Pending | | | |
| 1937 | Patent 2032695 * | | | |
| 1950 | Patent 2517191 | | | |
| 1957 | Full stamp with patent pending | | •••• | •••• |
| 1958 | Full stamp, no patent pending •••• | •••• | •••• | ••• |
| 1959 | •••• | ••• | ••• | ••• |
| 1960 | ••• | ••• | ••• | •• |
| 1961 | ••• | •• | •• | •• |
| 1962 | •• | •• | •• | • |
| 1963 | •• | • | • | • |
| 1964 | • | • | • | |
| 1965 | • | | | |
| 1966 | IIII | IIII | IIII | IIII |
| 1967 | IIII | III | IIII | III |
| 1968 | III | III | III | III |
| 1969 | III | II | III | II |
| 1970 | II | II | II | II |
| 1971 | II | I | II | I |
| 1972 | I | I | I | I |
| 1973 | I | | I | |
| 1974 | //// | //// | //// | //// |
| 1975 | //// | /// | //// | /// |
| 1976 | /// | /// | /// | /// |
| 1977 | /// | // | /// | // |
| 1978 | // | // | // | // |
| 1979 | / | // | // | / |

### ZIPPO LIGHTER IDENTIFICATION CODES

| YEAR | REGULAR LEFT | REGULAR RIGHT | SLIM LEFT | SLIM RIGHT |
|---|---|---|---|---|
| 1980 | / | / | / | / |
| 1981 | / | | / | |
| 1982 | \\\\ | \\\\ | \\\\ | \\\\ |
| 1983 | \\\\ | \\\ | \\\\ | \\\ |
| 1984 | \\\ | \\\ | \\\ | \\\ |
| 1985 | \\\ | \\ | \\ | \\ |
| 1986 | \\ | \\ | \\ | \\ |

EFFECTIVE 7-1-86 THE ABOVE SYSTEM WAS REPLACED BY YEAR/LOT CODE. YEAR IS NOTED WITH ROMAN NUMERAL/ LETTER DESIGNATES LOT MONTH (A=JAN., B=FEB. etc.)

| YEAR | REGULAR LEFT | REGULAR RIGHT | SLIM |
|---|---|---|---|
| 1986 | A to L | II | SAME AS REGULAR |
| 1987 | A to L | III | SAME AS REGULAR |
| 1988 | A to L | IV | SAME AS REGULAR |
| 1989 | A to L | V | SAME AS REGULAR |
| 1990 | A to L | VI | SAME AS REGULAR |
| 1991 | A to L | VII | SAME AS REGULAR |
| 1992 | A to L | VIII | SAME AS REGULAR |
| 1993 | A to L | IX | SAME AS REGULAR |
| 1994 | A to L | X | SAME AS REGULAR |
| 1995 | A to L | XI | SAME AS REGULAR |
| 1996 | A to L | XII | SAME AS REGULAR |
| 1997 | A to L | XIII | SAME AS REGULAR |
| 1998 | A to L | XIV | SAME AS REGULAR |
| 1999 | A to L | XV | SAME AS REGULAR |
| 2000 | A to L | XVI | SAME AS REGULAR |

**ZIPPO**, 1953. A lighter for the Phillies baseball team. $100-$275 (highest in Philadelphia)

**ZIPPO**, 1953. Zippo display with various lighters. $100-$125 (display card w/o lighters)

**ZIPPO,** 1953 & 1966. Club Crackers and Sunkist advertising lighters. The Club Crackers lighter is from 1953. $75-$100 each.

**ZIPPO,** 1955 & 1957. The man in his underwear is probably the closest thing to scantily clad models you will find. In the realm of ships, crossing the international date line is always celebrated with some ceremony to appease the gods of the seas. $75-$100

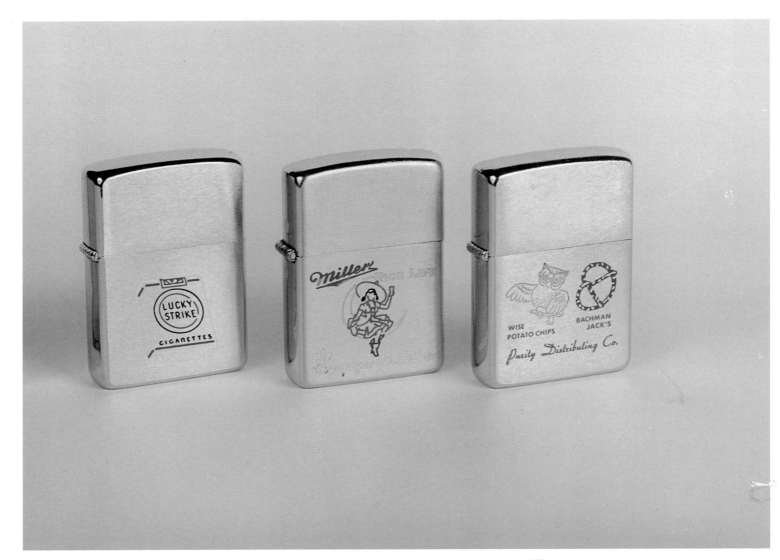

**ZIPPO,** 1953 & 1964. Three advertising lighters. The Miller Beer lighter is from 1953. $75-$125 each.

**ZIPPO**, 1958. Prototype "Town and Country" designs. $400-$500 each.

**ZIPPO**, 1958. Prototype "Town and Country" lighters in the design stages. $400-$500 each.

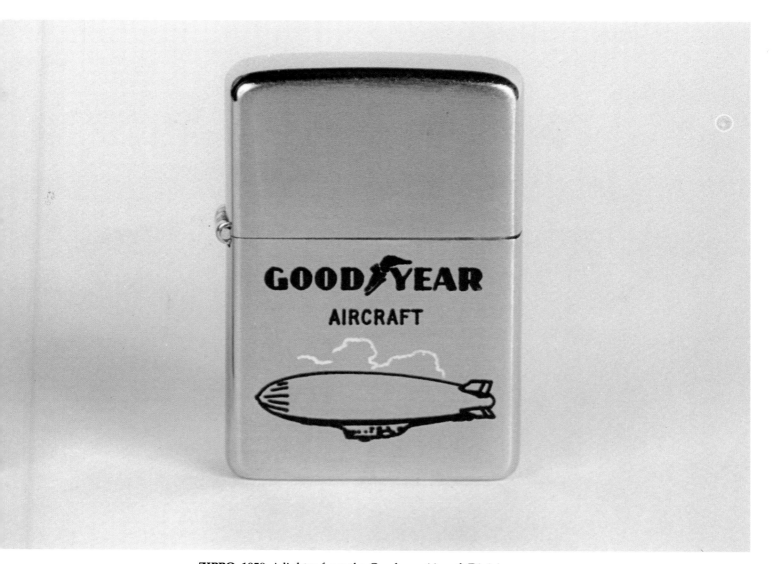

**ZIPPO**, 1958. A lighter from the Goodyear Aircraft Division. $125-$175

**ZIPPO**, 1958. Bell Aircraft. $100-$125

**ZIPPO**, 1958. Saturday Evening Post. $50-$65

**ZIPPO**, 1958. At first glance, this lighter does not seem very unusual. However, in light of the evidence linking smoking with heart disease and cancer, it is unlikely that you will ever again see a medical insurer advertised on a cigarette lighter. $50-$75

**ZIPPO,** 1958 & 1960. These lighters are often tougher to find than the astronaut mission (Apollo 11, etc.) lighters. The Aircraft Control lighter is from 1958 and the Cape Canaveral Lighter is from 1960. Cape Canaveral soon became Cape Kennedy. $80-$100 each.

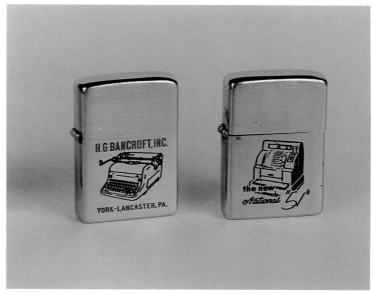

**ZIPPO,** 1958 & 1960. Two nice advertising lighters. The typewriter is earlier. $50-$75 each.

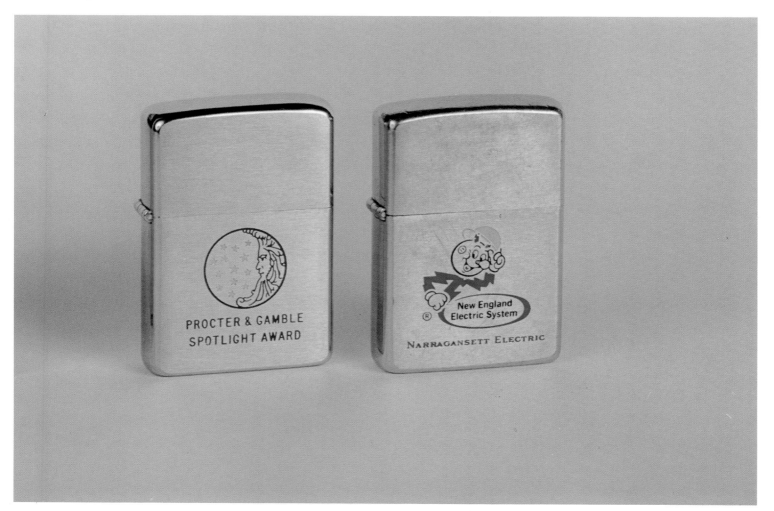

**ZIPPO,** 1958 & 1969. Proctor and Gamble Award (1958) and Naragansett Electric (1969) advertising lighters. L-R $80-$120, $60-$85

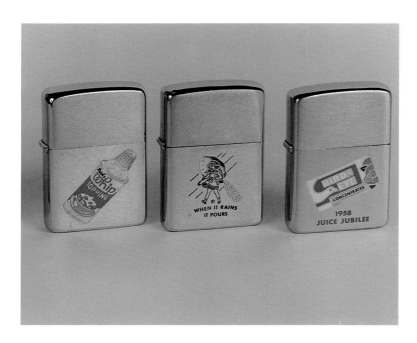

**ZIPPO,** 1958 & 1961. Three great advertising lighters for consumer products. The Morton Salt is from 1961. $85-$120 each.

**ZIPPO,** 1959. St. Petersburg Town & Country enamel model. $275-$300

**ZIPPO,** 1959. Whirlpool Home Appliances lighters (pocket and table). L-R $85-$100, $150-$175

**ZIPPO,** 1960. Zippo display case with various lighters. $100-$125 (case w/o lighters)

**ZIPPO,** 1960. This is an interesting aircraft/aircraft carrier lighter with the landing controller's paddle used to signal the pilot as he lands on the carrier. $150-$175

**ZIPPO,** ca. 1960. Zippo postcard from Bradford, Pennsylvania.

**ZIPPO,** 1960 & 1970. Coca Cola and Holiday Inn advertising lighters. The Coca Cola lighter is from 1960. L-R $185-$220, $125-$150

**ZIPPO,** 1962. Zippo catalog showing various lighters. Catalogs and display material are very difficult to find.

**ZIPPO,** 1964 & 1966. For the automotive collector. Bay State Gasoline and Willard Batteries. The Willard lighter is the earlier. Willard $50-$70, Bay State $75-$125

**ZIPPO,** 1967 -1981. Three colorful lighters for USS F.D. Roosevelt (1976), Recruit Training Command (1967) and USS California (1981). $40-$65 each.

**ZIPPO,** 1960s. Table lighters with advertising. L-R $75-$125, $150-$190, $150-$190

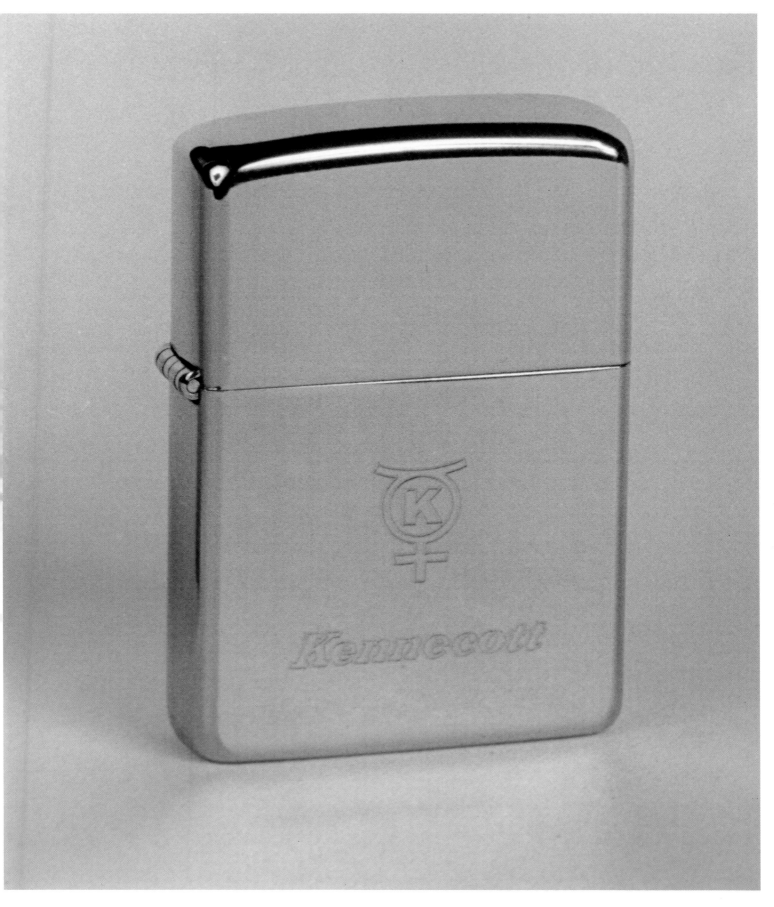

**ZIPPO,** 1969. A rare, all copper lighter made for the Kennecott Copper Company. $350-$450

**ZIPPO,** 1969 & 1976. On the right is one of the harder Apollo 11 lighters to find, along with a NASA Apollo program lighter. It is unusual that the Apollo program lighter has 1976 markings since the last Apollo flight to the moon was in 1972 and the Apollo-Soyuz flight was in July, 1975. Virtually nothing happened in 1976. $125-$175 each.

**ZIPPO,** 1970 & 1979. Campbell's and Gardner's advertising lighters. The Gardner is from 1970. L-R $75-$100, $100-$145

**ZIPPO,** ca.1970 & 1973. Two nice lighters advertising Zippo. Very difficult to find. $175-$225 each.

**ZIPPO,** 1973. A prototype design never commercially produced. $250-$325

**ZIPPO**, 1974 & 1982. Piels Beer (both sides) and Lucky Strikes lighter (both sides). $50-$85 each.

**ZIPPO**, ca.1976. The classic Mickey Mouse. $175-$225

**ZIPPO**, 1977. U.S.S. Theodore Roosevelt. A great image of Teddy Roosevelt holding a submarine. $125-$150

**ZIPPO**, ca.1976. A 1976 Zippo featuring the greatest race horse of all time. $75-$110

**ZIPPO**, 1980. Football team: The Tampa Bay Buccaneers. $50-$75

**ZIPPO**, 1981. Three prototype designs. One of the clues to defining these as prototypes (besides the excellent provenance) is the lack of a copyright symbol that would surely appear on a commercial product. $275-$325 each.

**ZIPPO**, 1981. A Popeye prototype design. $275-$325 each.

**ZIPPO**, 1982. Kool Aid lighter. $100-$125

**ZIPPO**, 1993. The Black Santa Claus is a rare number. It was made for foreign distribution. $150-$175

# BIBLIOGRAPHY

Balfour, Michael. *Alfred Dunhill: 100 Years & More.* London: Weidenfeld & Nicolson, London, 1994.

Cummings, U.K. *Ronson - The World's Greatest Lighter.* Palo Alto, CA: Bird Dog Books, 1992.

Bisconcini, Stephano. *Lighters/Accendini.* Milano, Italy: Edizioni San Gottardo, 1983.

Van Weert & Bromet. *The Legend of the Lighter.* New York: Abbeville Press, 1995.

# RESOURCES

Vintage Lighters, Inc., P.O. Box 1325, River Road Sta., Fairlawn, New Jersey 07410. Dealers in vintage lighters.

Authorized Repair Service, 30 W. 57th Street, New York, New York 10019. Dealers in vintage lighters and does lighter repair.

International Lighter Collectors Club, 136 Circle Drive, Quipman, Texas 75783. The largest lighter club. Publishes "OTLS" On The *Lighter Side* newsletter. Has information on the lighter shows.

Pocket Lighter Preservation Guild, 11220 West Florissant, Ste. 400, Florissant, Missouri 63033. Lighter club - keep trying, they have difficulty answering their mail.